GOAT MILK MAGIC

One of Life's Greatest Healing Foods

by

Bernard Jensen, Ph.D.
Clinical Nutritionist

PUBLISHED BY:

Bernard Jensen, Ph.D.
24360 Old Wagon Road
Escondido, CA 92027 USA

The health procedures in this book are based on the training, personal experiences and research of the author. Because each person and situation is unique, the publisher urges the reader to check with a qualified health professional before using any procedure where there is any question as to its appropriateness.

Because there is always some risk involved, the author/publisher are not responsible for any adverse effects or consequences resulting from the use of any of the suggestions, preparations or procedures in this book. Please do not use the book if you are unwilling to assume the risk. Feel free to consult a physician or other qualified health professional. It is a sign of wisdom, not cowardice, to seek a second or third opinion.

First Edition

BERNARD JENSEN, Publisher
24360 Old Wagon Road
Escondido, CA 92027 USA

ISBN 0-932615-32-5

CONTENTS

Dedication

Dedicated to those people who are looking for the best when they are sick. Without a doubt, they will find the best in the magic of goat milk.

Dedicated also to Meta Schuette, who lived at my Ranch for 30 years on goat milk and very little other food. To me, Meta was the greatest example of what goat milk can do for you.

PREFACE

Our goat herd at the Ranch is a very popular attraction to visitors and patients alike. The goat pen and shed are built on the kind of habitat goats love—a boulder-strewn sunny hillside with oak trees overshadowing part of it. The goats sometimes leap gracefully and playfully from boulder to boulder, entertaining all who watch with their antics, demonstrating the kind of agility and sure-footed balance their breed is famous for. No matter how old a goat is, he or she seems young.

Those who taste fresh goat milk for the first time at my Ranch are pleasantly surprised at how good it tastes. It is the closest to human milk of any milk I know, tolerated well by all ages, from babies to the elderly. I have seldom encountered an allergic reaction to goat milk as I have so often with cow milk. It is easily digested and assimilated, and is rich in the minerals and trace elements we need to prevent disease and enjoy the best of health.

WHY PEOPLE HAVE GOATS

I made it my business to visit farms and rural homes that had goats, wherever I found them. I always asked why they had the goats. The reply was always, because of health problems. I never found any place where they used goats for any other purpose.

A farmer might say, "My mother had tuberculosis, and goat milk helped her." A housewife would tell me, "My husband had stomach ulcers, and he doesn't have them any longer because of using the goat milk." Another said, "In weaning my baby, I couldn't find any substitute for mother's milk until we found goat milk."

Of late, I had a patient who was allergic to cow milk but not to goat milk. This girl was lacking in calcium, and her joints were poor. I mentioned she needed milk for its wonderful calcium and I advised using goat milk. While she had been unable to take cow milk, she was immediately able to take the goat milk. This proved to be a boon to her because her whole body changed, and she is getting the calcium she needs from one of the finest sources I know.

PREVENTION IS BETTER THAN CURE

It was Professor Max Rosenheim who wrote in the British medical magazine *Lancet*, "It must increasingly be the purpose of the medical profession, and of all who work with it, to aim at prevention rather than cure...." *Lancet* carried the following by J.N. Morris: "To improve health, lift the burdens of chronic disease, and reduce the death rate, the main hope lies surely in prevention." I honestly believe that goat milk is one of the most nutritious foods known to man. If used more widely, it could be one of the best ways we have of preventing disease.

Over the 50 years of my work in the healing art, I have seen seemingly hopeless cases revived from the death-bed with fresh, warm goat milk. My own mother got rid of her tuberculosis after using goat milk. Milk fresh from the goat carries the life energy with it. When it is taken warm, the greatest benefit is received, since it is immediately digested at body temperature. Goat milk digests in 20 minutes, in comparison with cow milk, which takes 2-to-3 hours to digest. Goat milk is nutritionally balanced to nourish a kid weighing 6-to-10 pounds, about the same as the weight of a human baby.

Cow milk is for a calf weighing 65 pounds at birth. The fat globules in cow milk are 5 times larger, which is why cow milk is much harder to digest than goat milk. It is also the reason why some babies get colic after drinking cow milk.

GOAT MILK—A SUPER NATURAL HEALTH DRINK

Goat milk promotes good health in many ways. One of its most remarkable features is that for young and old alike who are prone to digestive distress of any kind, goat milk is the one food that stays down better than all other foods. The most sensitive stomach seems to welcome goat milk. The most fatigued body seems to revive and thrive on it.

I use goat milk. Many of my employees at the Ranch over the years have used goat milk. The great majority of my patients have chosen to include goat milk in their diets, when it was an option. I sometimes insist that certain patients use goat milk because I know they will not improve without it.

I may have a glass of goat milk a day, half a glass mixed with carrot juice. When I'm busy working at the Ranch and need a pick-me-up, I just say, "Give me my drink," because I don't have time to sit through a meal. My drink is made up of raw goat milk, a tablespoonful of sesame seed butter (or almond nut butter), a teaspoon of honey and a sliver of avocado or half a banana. At times I may ask for a raw egg yolk to be added to it. This makes a very nutritious drink.

Some years ago, the Journal of the *American Medical Association* published an article titled "Dietetics and Hygiene," which reported: "The goat is the healthiest domestic animal known. Goat milk is superior in every respect to cow's milk. Goat milk is the ideal food for babies, convalescents and invalids, especially those with weakened digestive powers. Goat milk is the purest, most healthful and most complete food known."

Goat milk is excellent for the sexual system and the glands, according to Dr. V.G. Rocine. It lifts those suffering from low vitality, and it is charged with sexual vitality.

The balance of protein, carbohydrates and fats in goat milk is excellent and a quart of raw goat milk is only 630 calories. The amount of calcium and phosphorus in a quart of goat milk is adequate to supply the average person's daily needs, and significant amounts of all the vitamins, minerals and trace elements needed for wellness. Goat cream is the easiest cream to take into the body. It is wonderful for putting weight on an underweight person. I'll discuss the nutritional benefits of goat milk at greater length in Chapter 6, "Super Nutrition of Goat Milk."

HOPE FOR BETTER HEALTH FOR THE AGING

The average life span for Americans is 72.5 years, according to the latest research figures. I've already outlived that statistic by more than 12 years. My family is not particularly long-lived in previous generations, so I feel that my food regimen and lifestyle have carried me far beyond the average life span. I believe people of all cultures can feel better and live longer if we eat wisely and take better care of ourselves.

Aging doesn't have to carry senility, arthritis, asthma and osteoporosis with it. I've met men in the range of 120 to 150 years old in the Caucasus mountains of Russia. Do you know what they were eating? Clabbered goat milk, raw honey and whole cereal grains.

I believe goat milk can help the generations live better as they live longer.

CONSIDER THE SOURCE

When I talk to my students and patients about the foods they should eat, the first thing I tell them is to consider the soil the food is grown in. We have to consider the source. Any particular food is only as good as the soil it is grown in. When we consider animal products such as milk, we have to consider both the food the animal is given and the animal itself.

The goat is the cleanest of the milk-producing animals and the most disease-resistant. This speaks well of the high quality of its milk. It is an intelligent animal, friendly, sensitive, gentle, patient and loving. It is virtually fearless. We have coyotes in our area, but they do not bother our goats because the goats can take care of themselves. There is no need for concern that goat milk will be contaminated with "fear acids" or metabolic chemicals produced by fear because goats are not timid, fearful animals. They do not "spook" easily, like cattle and sheep tend to do. Goat milk is an ideal food for lacto-ova vegetarians for this reason. There is no other food from an animal source that comes from a more stable, peaceful, mental and emotional background than that of goats.

The cleanliness of goats is remarkable. There are few flies around a goat pen, because the pellets excreted by goats are small enough to dry quickly, and very little odor is produced. Cattle and chickens attract far more flies than goats.

Goats are resistant to tuberculosis, a disease that cattle are very prone to. My mother once had tuberculosis and got rid of it by using goat milk. I have had very few health problems with my goats, summer or winter. Only rarely do I have a goat examined or treated by a veterinarian. When this happens, it is more likely to be from an injury than a sickness.

Pigs will eat anything given to them, but goats are very picky. I have seen my goats refuse an apple that someone has taken a bite out of. Goats will nibble on the leaves of trees, bushes, shrubs and chaparral. They will pull strips of bark from a tree and eat it. They are not grazers like cattle, but browsers—very selective in what they eat.

As a result of a more varied diet, as well as nibbling at the leaves and bark of deep-rooted trees and bushes, goats take in more trace elements and a greater variety of nutrients in general than almost any other livestock. For this reason, goat milk is probably the most nutritious milk available in any particular environment or location.

Dr. Rocine stressed that we should take milk only from a young and healthy animal with lively joints, because the milk is representative of the whole body of that animal. A young animal gives the greatest amount of sodium in its milk. I might add that I have seen the sick practically raised from the deathbed by the use of warm raw goat milk. Goats are a part of my life due to the wonderful work I have done with raw goat milk. It has been important in my own health, and comes from a very limber animal, a sodium animal.

WHY I DRINK RAW GOAT MILK

I believe the human body was designed to be nourished and kept going by foods (with some exceptions) as nature made them. Pasteurization of milk not only kills germ life but destroys enzymes and changes the chemical balance of the milk, so it is not the same as nature made it. I believe a calf raised on pasteurized milk would become sick over a period of time, if it survived at all. In fact, my opinion that raw milk is superior to pasteurized milk is based on scientific evidence.

In the 1930s, I had the privilege of visiting Randleigh Farm in Lockport, New York, the most scientific and advanced farm of its kind in the country at that time. The owner, William R. Kenan, Jr., conducted a series of experiments in 1941 that compared the health-building qualities of raw milk to those of pasteurized milk and boiled milk. (Pasteurized milk is not boiled, but heated to 140-150 degrees F. and kept there for 30 minutes, then quickly cooled.) Groups of laboratory rats were put on a diet of the different kinds of milk supplemented with minerals for four-and-a-half months, while control groups were fed regular commercial rat food. The examination of the rats was done by a medical doctor, Dr. Francis M. Pottenger, Jr.

Keeping in mind that milk is not a natural diet for rats, the raw milk-fed rats were in much better condition than either the pasteurized milk-fed rats or the boiled milk-fed rats. The 27 rats in the raw-milk fed group were generally healthy, with lung lesions in two rats, a kidney lesion in one and mottled livers in three. Of the 27 rats in the pasteurized milk group, two showed evidence of heart injury, five had kidney lesions, 3 had gross liver conditions, five had severe lung pathology and three others had pulmonary disturbances. There were six lesions in the 14 boiled milk-fed rats, four lung lesions and two liver disturbances. Dr. Pottenger also found that the livers were smaller, the bones stronger and the muscle tone was better in the raw milk-fed rats as compared to the others.

This experiment was done with cow milk, but I believe the same superiority of raw goat milk over pasteurized goat milk would have been shown if it had been done with goat milk.

Raw goat milk builds the brain and nervous system. When we provide the cholesterol, phosphorus, amino acids, silicon, sulfur and vitamins needed by the brain and nerves, we are not only feeding the brain but helping the whole body. The brain serves as the orchestra conductor for the whole body, and when the brain is taken care of, it takes better care of every organ, gland and tissue in the body.

I believe in raw goat milk for many other reasons. I know it has kept me healthier and stronger than I would have been if I hadn't used it. I've seen it help bring thousands of men, women and children back to good health. I feel there is solid evidence that it increases longevity, and that many, many lives have been healthier and happier because of it.

"Here's lookin' at you, kid!"

INTRODUCTION

ALL ABOUT GOATS

I feel that goats are among the most valuable and fascinating of God's creatures. I visited several hundred goat farms and dairies before making the decision to have goats myself. I've kept small herds of French Alpine milk goats at my sanitariums since 1941 because the fresh goat milk has proven itself a valuable aid to recovery of my patients from many diseases and a valuable source of added strength and comfort to the permanently disabled or to invalids. Some doctors call goat milk "prescription milk," because it is so widely prescribed by doctors and nutritionists for everything from infant colic to advanced degenerative diseases.

Goat milk has saved countless people from big hospital bills by preventing diseases and disturbances that would normally have required hospital care. Goat milk has saved thousands of invalids from having to be confined to convalescent homes. Fresh goat milk strengthens many invalids enough to enable them to take care of themselves and stay in their own homes. I feel that goat milk has saved countless weak babies and critically ill patients from an early grave. I've seen this happen with my own eyes. Goats are invaluable creatures, and goat milk is a wonderful blessing to mankind.

Cartoons showing goats eating tin cans, rubber tires or clothes off clotheslines are misleading. Goats have very sensitive digestive systems and are very selective about what they eat.

I want to make clear from the beginning that not all goat milk is magically health-building. You can't starve a goat and get good milk out of her. You can't put her in a patch of weeds and thistles, or give her your kitchen garbage, and expect milk that's worth much. The nutritional value of goat milk is directly related to the nutritional value of what goats are fed, and if you want milk that is high in nutritional value, you must feed your goats the best feed you can get.

When they have a choice, goats will refuse poor grades of alfalfa and eat only the high-quality alfalfa. Given a choice between a depleted, poorly maintained pasture and a well-fertilized high-grade pasture, they'll pick the best pasture every time.

Goats are not only highly-intelligent animals, they are loving and friendly. They respond with affection to kind treatment and they will give more milk to milkers they like than to milkers who are indifferent toward them. If we look back in history, we find out that goats and mankind have had a long and valuable friendship.

GOATS IN ANCIENT TIMES

When the Bible refers to Palestine as "the land of milk and honey," the milk it mentions is goat milk. According to archaeologists, the first evidence of man using domestic goats was in Southwest Asia around 7,000 BC. This is the region where nomadic tribes learned to tame and keep wild goats for their milk and meat.

In the story of the prodigal son in the Bible, his jealous brother complains to their father, "You never so much as gave me a young goat, so I could make merry with my friends." Beside using goats for food, goatskin tents were used to protect families from the desert heat, cold and sandstorms. Goatskin containers were used for water bags, or to carry milk.

The first "curds and whey" were probably produced when milk carried in a water bag made from a goat stomach became mixed with a little rennin and produced the first cheese. Goat cheese is popular in many countries of the world, even today. Clabbered milk, yogurt and several kinds of cheese can be made from goat milk.

Methuselah, the oldest man in the Bible at 969 years, may well have reached that very ripe old age with the help of goat milk and products like clabbered milk. He lived in the area where wild goats were first tamed for their milk and meat. Methuselah may have used goat milk along with dates, figs, grapes, honey, barley, millet, lentils, onions, leeks and garlic. These are the foods documented in the ancient writings for that area of the world.

When I visited the Caucasus mountains of Russia years ago, I found some of the oldest men in the world, ranging from 120 to 150 years of age. Among their foods were clabbered goat milk and raw honey.

Goat milk was one of the foods the Greek physician Hippocrates recommended to his patients. He is credited with saying, "If you want to be well, get a goat and live on the south side of a mountain." Hippocrates, called the father of the healing arts, was born in 460 BC and lived till the respectable age of 84.

Although goats were originally wild mountain dwellers, there are now over 200 breeds and varieties in the plains, valleys and deserts of the world. Goats are everywhere!

GOATS ARE WORLD TRAVELERS

I don't care where you go, you will find goats. You will find goats in Siberia. You will find goats in Pategonia. Goats are as much at home near Mt. Fiji in Japan as they are in the Alps of Switzerland. They love the green valleys of Chile and the lowlands of the Philippines and Nigeria. Goats seem to be happy everywhere they are taken, as long as they have food they like and people who care for them.

India takes the prize for the highest populations of goats with 65 million. China is number two with 55 million goats. Nigeria and Turkey are tied with 20 million each, and Iron is not far behind with 18 million goats. They are valued for their meat, milk, cheese, hides and hair (from the angora and cashmere breeds).

For those interested in goat trivia, one difference between goats and sheep is that goats hold their tails up, while sheep tails hang down. Male goats are called bucks (or billies by city folk) while male sheep are rams. Female goats are does (nannies to city folk), and female sheep are ewes. Most bucks are bearded while does are not, except the Kasmiri doe.

In the country of India, goats are trained to be work animals. Milk and cheese are the main interests in Western Europe, while in Turkey and Central Asia, angora and cashmere goat or hair are made into yarn and then into very popular clothing. Goat milk is used in virtually every country of the world.

Nearly all of today's breeds and varieties of goats are descended from the wild bezoar or the wild goat of Pakistan, even the most popular breeds of the U.S.

HOME, HOME ON THE RANGE

It isn't only deer and antelope that play on the rangeland of the united states. There are four million goats in the United States, nearly all used for milk, cheese and other milk products.

The most popular milk goats in the U.S. are the Nubian and the Alpine: American and French Alpine. Saanens and Toggenburgs are also European breeds, but they are not related to Alpines. The other two dairy breeds found in the U.S. are LaManchas and Oberhaslis. The Nubian we have in this country was originally from Africa but was later interbred with goats from India and Great Britain. The French Alpine was developed in the French Alps. The Saanen, one of the biggest breeds, is from the Saanen Valley in Switzerland. Another Swiss goat is the Toggenburg, from the Toggenburg Valley. This is a small goat.

When I say "large" and "small" in reference to goats, I mean that goats range from the Pygmy goat, a foot-and-a-half high and 40 pounds in weight to the Saanens, over three feet high and up to 250 pounds. The females are considerably smaller than the males.

Pygmy goats are popular in this country. So are angoras. The U.S. is the biggest producer of angora wool in the world, producing over 25 million pounds a year. The fleece of the angora is actually called mohair. Kashmiri goats are apparently not found much in the U.S., since the big cashmere production (the hair and the yarn) is out of Asia.

Raw goat milk is a specialized industry in the U.S., usually produced in the cleanest dairies you can find. The rules and regulations for the production of raw goat milk are sometimes almost unbelievable. But, there are consumers who believe that raw goat milk is the best, most nutritious health-building milk in the world, and they won't have anything to do with pasteurized goat milk. I agree with them. I believe in raw goat milk.

GOAT NUTRITION

I want to give some reasons why goat milk is so highly regarded by health professionals, and in doing that, I need to provide some background on goat breeding and milk production. It should be apparent that does give milk only after giving birth to kids—usually one to three. After the kid or kids are weaned from the mother, the milk cycle begins and lasts about 10 months. A young doe is usually bred at about a year-and-a-half and gives birth five months later. The second time she is bred, she can still be milked for about three-and-a-half months into her pregnancy. This period is a critical nutrition time for the doe, because of the nutrients required by one or more fetuses and the milk nutrient level. The doe must be fed well.

The critical nutrients before breeding and during pregnancy are adequate carbohydrates to keep the blood sugar up, phosphorus, calcium, manganese, sodium, potassium, chlorine and trace minerals. Vitamin A and pro-vitamin A (carotene) are both very important for healthy reproduction. Vitamin D is needed to allow the calcium and phosphorus to be utilized properly. Vitamin E and selenium are believed to protect and enhance the health of the mother goat prior to and after conception.

I feel it is necessary to point out that, like mineral-deficient, stressed-out people, mineral-deficient stressed-out goats can get sick or develop diseases. Goats don't get tuberculosis or brucellosis like cows often do, but they get other things. Goats can take heat or cold, but they don't tolerate wind, drafts or chilly rain well. They need clean water. They don't do well in a noisy, high traffic area (either people or machinery). Under stress and with poor feed, goats may get pneumonia, colds, worms, clostridia and shipping fever. Worms, of course, are due to environmental conditions, and goats have to be wormed several times a year in some places.

A milk doe, on the average, gives two quarts of milk a day. A lot of energy and nutrients are required to produce that amount of milk and these must be replaced for the doe to give good, high-nutrient milk day after day.

Goats tend to refuse feed they don't like, and their milk level drops. In New Zealand, grass-clover pasturage satisfies dairy goat herds, and high-quality milk is produced. In some parts of the world and this country, goats are not satisfied with available forage, and the milk quality and quantity are lowered if they don't get supplementary feed and nutrients.

A good goat feed for milk production is high in the same minerals and vitamins needed before breeding and during pregnancy, with the addition of magnesium and sulfur. Vitamins A, D and E continue to be important. Hay given in addition to forage increases milk production and quality.

In addition to good feed, goats need a clean pen (they are clean themselves—much cleaner than cattle or pigs), adequate shelter from wind, keepers who recognize symptoms of diseases and mineral deficiencies and know what to do about them), adequate water and, ideally, rocks, logs and other things to climb on to keep their hooves trimmed. Goats like to get up high on things, possibly due to their mountain heritage.

Given good food, habitat and care, goats are unusually peaceful, contented and stable. They don't spook like cattle, horses and sheep. They are not intimidated by predators like coyotes and wolves, and mountain lions will seldom attack a full-grown goat, although they may carry off a kid. The calm, placid emotional stance of goats tends to favor the efficient conversion of a high-nutrient feed into high-nutrient milk—and this is what we are looking for.

The high-nutritional quality and the greater ease of digestion of raw goat milk, taken fresh from the goat and still warm with life energy, is what makes goat milk so effective in the prevention or reversal of disease. The secret is that it is naturally half cleansing and half building.

There are over 375 million goats worldwide. Isn't that wonderful? Maybe we need more. I feel everyone in the world should have access to fresh goat milk.

Dr. Jensen, nature student.

Dr. Jensen seated at the Ranch's Hippocrates Circle, where he teaches the value of foods and nutrition.

CHAPTER 1

GOAT MILK—MORE THAN A FOOD

I once had a patient from Seattle who was six feet three inches tall and who couldn't have weighed more than 115 pounds or so. He was just skin and bones, as they say. His doctors couldn't find anything wrong with him, but he had lost nearly 65 pounds, so his family sent him to me. I knew he was coming and I had given considerable thought to what I was going to do.

When he first arrived, we had a conference in my office. "Dr. Jensen," he said, "I came down here to see you, because no one back home knew how to help me. I'm ready to do anything you tell me to do." I want to tell you, a doctor likes an attitude like that. This man came to me believing I could help him. An attitude like that is a great help in bringing a patient out of the old life and in moving him toward wellness.

"Don't be surprised at what I'm going to tell you," I said. "I have a herd of French Alpine goats on the Ranch to provide milk for my patients. I want you to learn to milk a goat, and when you've milked enough to fill a big glass, I want you to stop and drink it. I want you to make friends with that goat and milk her 7 or 8 times a day. Drink the milk, warm and foaming, so you get the animal magnetism and life energy straight from the goat."

The vibratory rate of the life energy in fresh raw milk is highest when it first comes from the udder. It diminishes little by little and is lost three hours after the goat is milked. I have seen raw goat milk practically pull people back from the dead.

In less than four months, that man gained 40 pounds. A positive attitude is a healing attitude, and when patients come to my Ranch, I always tell them they are going to have to work to get well. Work helps to reinforce a positive attitude.

After six months, this patient had gained 50 pounds, and we decided it was time for him to go back home to Seattle. When I saw him off, his car had a trailer behind it, and in that trailer he had a goat.

THE SUPERIORITY OF GOAT MILK

Although cow milk and dairy products are popular in the United States and Western Europe where an abundance of alfalfa and supplementary grain feed are available, they are not so popular in the rest of the world. A survey not long ago showed that 65% of the milk used in the world is from goats. There are also places where mare's milk, camel's milk and sheep's milk are used. When both are fresh, the flavor of goat milk and cow milk are comparable. The nutrient content of goat milk is similar to that of cow milk, but goat milk is more digestible, because the fat particles are smaller.

The late Mahatma Gandhi of India has been an inspiration in my life, and I feel that I've shown my gratitude to him by being partly responsible for his decision to include goat milk in his food regimen.

It was Gandhi's personal decision to fast whenever there was a major political confrontation or disagreement between the Indians wanting independence and the British rulers. Gandhi led India to independence by the influence of his peaceful, non-violent methods of political appeal for justice and self rule.

But, he fasted so much—once as long as 21 days—that his health was affected. He couldn't regain his weight on a vegetarian diet. So, I wrote a letter suggesting that goat milk might be helpful in restoring his normal weight. He didn't answer my letter, but I noticed in the newspaper that the next time he traveled to England, he took a milk goat along on the ocean liner. I was pleased that he had tried my advice and found it gave good results.

Goats, generally speaking, are more disease resistant and economical to keep than cows. This means a lot in Third World countries where veterinarians are scarce and supplementary feed is often unavailable or unaffordable to the average family. But, even in Western countries, the lower cost and greater convenience of keeping goats is an attractive feature. Goats are browsers, not grazers, and they make better use of available vegetation.

I believe goat milk is healthier than cow milk. I've seen many adverse reactions to cow milk and very few with goat milk. It isn't uncommon to find people who have had trouble with cow milk switching to goat milk and doing very well. I often recommend that people with allergy problems or catarrhal problems try using goat milk instead of cow milk.

The following comparison of nutrients in goat milk, cow milk and mother's milk is helpful in seeing the value of goat milk. However, because goat milk is low in folic acid, it is a good idea to take a 100 microgram supplement of this vitamin for every quart of goat milk.

THE CHOLESTEROL QUESTION

Many people contend that adults should use only low-fat or no-fat milk because of the cholesterol content of whole goat and cow milk. I believe whole foods are better for us, because the Creator knew exactly what He was doing when He put certain groups of vitamins, minerals and other nutrients together. Now we find out there is "good cholesterol" (HDL) and "bad cholesterol" (LDL and VLDL), and that even the bad cholesterol is only bad when there is too much of it. We need a certain amount of both.

Because the thyroid is higher evolved in the goat, the fat content of the milk is very well controlled. That is one reason why the fat globules in goat milk are one-fifth the size of the fat globules in cow milk. There is no cream line, no separation of the cream from the milk because of the smaller fat globules. Further, there is not as much pro-vitamin A in goat milk as iin cow milk. There is more of what we call pre-formed vitamin A, so it can be immediately used by the body. The vitamin A in goat milk is carried in the fat globules.

The cholesterol in goat milk is more highly evolved than that of cow milk and is more useful to the brain and body. The brain uses cholesterol to form and maintain the myelin sheaths of the nerves in the "white matter" part of the brain. Cholesterol and lecithin are used by every cell of the body to transport nutrients through the cell membrane, and a certain amount of both are constantly circulating in the blood. When we have enough lecithin, cholesterol is kept in solution in the blood. Only when we overheat or overcook a food is the lecithin destroyed, allowing the cholesterol to remain. When such a food is digested and assimilated, the out-of-balance cholesterol is able to be deposited on the walls of the cardiac and pulmonary arteries. The brain, nervous system, glands and cells use up a great deal of the cholesterol manufactured by the liver and taken in as food, but a surplus of cholesterol leads to atherosclerosis.

So, what is my verdict? I believe the benefits of whole, raw goat milk outweigh any risks that the cholesterol may pose. A quart of whole goat milk may have as much as 112 milligrams of cholesterol. In an overall, balanced diet, deliberately holding back on concentrated fats and oils, there is no problem with the amount of cholesterol in goat milk.

It is important to drink milk when it is fresh and warm and full of the life energy of the goat, because, at first, it is alkaline and tends to neutralize acids in the body. If it sits, the lactose breaks down into lactic acid and the milk becomes acidic. This is not as good for us as the fresh milk.

COMPARISON OF NUTRIENTS IN MILK

Item All Items Per Liter		Breast Milk	Goat Milk	Cow Milk
Energy	Kcal	710	670	660
Protein	gm	11	32	42
Fat	gm	38	40	37
Carbohydrate	gm	68	46	49
Calcium	mg	340	1290	1430
Phosphorus	mg	140	1060	1120
Sodium	meq	7	15	27
Potassium	meq	13	46	45
Iron	mg	5	1	0.5
Zinc	mg	3.5	2.4	3.5
Chloride	mg	375-450	1200	1050
Magnesium	mg	46	100-145	120
Vitamins				
A	IU	2000	2074	1500
B-1	mg	0.160	0.400	0.440
B-2	mg	0.360	0.1840	0.2100
B-3 (niacin)	mg	1.47	1.9	1.0
B-6	mg	0.100	0.70	0.640
B-12	mcg	0.3	0.6	4.3
Pantothenate	mg	1.84	3.4	3.5
Folacin	mcg	52	6	55
Diotin	mcg	8	39	31
Choline	mg	90	150	121
Inositol	mg	330	210	110
C	mg	43	15	21
D	IU	22	24	14*
E	IU	1.8	—	0.4
Essential fatty acids, per 100 gm of milk fat			4.1	2.6
Percent fat globules less than 3 dram			63	43

*Not fortified

THE POOR MAN'S MEDICINE CHEST—A GOAT

I don't know how many times I've driven down country roads, spotted a milk goat in a pen or barnyard, and stopped to see why the people had it. I heard some wonderful testimonies.

There was Uncle Lester, who ruined his digestion at age 18 and had to drink goat milk the rest of his life. He was 93 when I met him. There were babies, whose mothers couldn't give breast milk. The babies thrived on goat milk. There were asthmatics of all ages, who suffered much less when they switched from cow milk to goat milk. Invalids nearly always did better on a diet containing fresh, warm goat milk, and in the days when tuberculosis was on the rampage in this country, goat milk strengthened those who had it better than anything else I know.

Malnutrition, lung and bronchial troubles, weak digestion, rickets, catarrhal problems, joint troubles and underweight problems—all improve when goat milk is used. Catarrh is an inflammation usually produced in the membranes of the body due to an irritant of some kind either from within the body or from the outside, as in the case of an allergy reaction. The body's response to catarrh is to produce mucus in the area of inflammation in an attempt to get rid of the cause. Catarrh marks the beginning of disease. I consider goat milk—raw, of course—a tonic for the elderly, the weak and young children, and a health builder and disease preventer for normal folks. Goat milk does not cause the formation of phlegm (catarrh), unlike cow milk.

For those recovering from devastating diseases, there is nothing like warm, foaming milk, straight from the goat to restore strength. The vital energy of the goat remains in the milk for about three hours. Then it is gone.

I have seen goats in Mexico and other Latin American countries, thriving on forage where a cow would starve to death. I have observed very fine herds of goats in Greece, France and Switzerland. Third World countries tend to use goats more than cattle, because they are hardier, less expensive to keep and provide more nutritional value than most other animals, except possibly chickens. (I consider the egg yolk of a boiled or poached egg, the best of all proteins.)

Goat milk has more chlorine, fluorine and silicon than any other milk from domestic livestock. This chlorine and fluorine make it naturally germicidal, and the silicon tends to help prevent diabetes. In rural areas with few doctors, it pays to prevent disease. Goat milk helps us keep cleaner, healthier bodies.

So many of the longevity champions I've met in my travels used clabbered goat milk in their diet. Clabbered milk is soured and coagulated at room temperature, which makes it more digestible while retaining the food value. I've met Russian men in the Caucasus mountains who were over 150 years of age.

A DOCTOR RESEARCHES GOAT MILK

Dr. Carl Wilson of Palo Alto, California, made a careful study of goat milk and developed a set of thoughtful reasons why goat milk is a nutritionally superior food for infants, invalids and ordinary people. The following are his conclusions:

- Goat milk is a far better emulsion than cow milk.
- The fat globules are 1/5th the size of those in cow milk.
- The reaction of goat milk is alkaline, the same as mother's milk. Cow milk produces an acid reaction.
- Goat milk is about 1/3 richer than cow milk.
- The curd in goat milk is small and light, hence easily digested. The curd in cow milk is large and dense.
- Goat milk is 2% curd (which precipitates in the stomach) as compared with 10% curd in cow milk.
- A goat is an exceptionally healthy animal, immune from tuberculosis, consequently gives healthful milk of uniform consistency.
- The goat is a very particular animal. It likes to eat only the top or honey leaves.

My own personal conclusion from having French Alpine milk goats for over 50 years, is that goat milk is a wonderful help in restoring the health and strength of invalids, elderly persons and many of those with chronic diseases. I've seen goat milk pull sickly babies through. I've seen it work as a survival food for people who can't take any other foods. I've seen goat milk put weight on anorexic patients. Goat milk is one of the best natural health supplements I know.

THE PROBLEM OF EXCESS

There is such a thing as "too much of a good thing." Today many people have gone overboard in their use of two products that are easy to produce in Western nations and that are used as ingredients in a great number of other foods. I'm referring to wheat and milk, and their many uses. In my book ***Vibrant Health from Your Kitchen***, I point out that the average American diet is 29% wheat and 25% milk. This if far too much!

Those who crowd out other foods to fill up on milk products and wheat products run the great risk of deficiencies that would not occur if they used a greater variety of foods. Children who use too much milk run the risk of anemia. Both cow milk and wheat can trigger catarrhal problems that create a "disease-friendly" environment in the body. Children raised on cow milk tend to be far more prone to catarrhal troubles than children raised on goat milk, since goat milk does not produce catarrh. Too many Americans are wheat-logged and milk-logged.

Looking at the beautiful variety in God's garden, 25% milk and 29% wheat is too much. The proportion is off balance. No more than 5% of the total diet should be consigned to milk and wheat. A 150-pound man with moderate activity may use 3,200 calories. Five percent of that amount is 160 calories. If we cut down to 5%, allergies, bronchial troubles, asthma, colds and flu would be under control. Doctors agree that we use too much milk and wheat. We need more greens, more salads, more of the vegetable kingdom, more fiber foods and fewer fast foods, packaged foods and processed foods.

A quart of goat milk is about 670 calories. But, there is more to a diet than calories, and we have to look at nutrient priorities together with our total caloric intake. The purpose of evaluating calories is to see if our energy needs can be met by the foods we eat. We still have to take in the right kinds of foods and nutrients to keep our bodies healthy and strong.

A quart of goat milk is not too much for a healthy teenager, provided that enough vegetables, fruits and whole grains are part of his overall food regimen. The point is we have to have as much variety as possible, a fat intake low enough to allow us to have a healthy heart, 60% raw foods, and we need to take greater care in our protein and starch intake. Think **B-A-L-A-N-C-E-D food regimen**, and it will all work out.

Each year kidding time sparks a lot of interest for employees and visitors to the Ranch.

Dr. Jensen with two new kids. Goat milk gives the sodium that is so necessary to keep man's joints supple, limber and pliable.

CHAPTER 2

DISCOVERING GOATS AND GOAT MILK

I discovered the healing value of foods the hard way. After working my way through college and graduating in 1929, my health broke down. I was diagnosed as having bronchiectasis, a lung infection for which there was no cure. My mother died young with a lung ailment, and I had inherited her lung weakness. Since there were no antibiotics in those days, my doctor told me to go to bed and rest. I wasn't about to risk repeating my mother's fate, and feeling very anxious and frustrated, I began to search for someone in the healing arts who could at least give me hope.

A Seventh Day Adventist doctor who knew the healing value of foods was the first one to put me on the right path. He pointed out the inadequacies of the junk food diet I'd been living on, and patiently explained to me the connection between a properly balanced food regimen and good health. As a result of his advice, I changed my food habits completely. Soon, I began to regain my strength and improve my health.

To be honest, I can't remember whether the Seventh Day Adventist doctor talked to me about goat milk or not, but I soon met a doctor who did. His name was V.G. Rocine.

ROCINE'S PHILOSOPHY OF HEALTH AND DISEASE

Dr. Rocine was a Norwegian homeopath who believed that since man was made of the dust of the earth, the mineral elements which make up our bodies were the keys to understanding health and disease. He believed that when we have all the minerals we need, we live in health. Dr. Rocine believed that excess or deficiency of minerals created the conditions for a disease.

I heard Dr. Rocine lecture on foods in the 1940s, and I was awed by his immense knowledge. He was decades ahead of his time. He had studied the works of the foremost European food chemists. He knew his stuff. He was a strong advocate of using goat milk, because of the broad array of minerals found in it and their benefit to human well-being.

In private conversation, Dr. Rocine convinced me that fresh, unpasteurized goat milk is an invaluable aid to the recovery of patients who may have many different diseases.

MY FIRST GOAT HERD

I began operating my second sanitorium in Altadena, California, in the 1940s, and it was in 1944 that I had my first small herd of milk goats. I can tell you, it is one thing to hear about the merits of fresh goat milk, and it is another thing to have the goats on hand to have easy access to their fresh milk. Goats are quite an adventure.

One of the first things I found out was that goats are smart enough to get out of almost any kind of enclosure or pen, until they decide that they want to live there. I don't think there is any latch that can be unlocked with human hands that can't also be unlocked by a goat with its hooves, teeth and lips. Once the goat is out, others are likely to follow. Then, some will return easily to the pen, while others dare you to catch them.

You can have a good laugh at some of the antics of your goats, and after 40 years of having goats, I have to say I enjoy being around them. But, they can be a trial to a person's patience. They always seem to get out just before you're on your way to keep an appointment. The more impatient you are to catch them and get them back into the pen, the more playful and evasive they seem to become.

I must stop here to say that it was all worth the trouble of learning the peculiar qualities of goat behavior to discover the world of good that fresh goat milk brought to my patients.

I've had patients who were brought to my sanitoriums in wheelchairs or on stretchers in ambulances. I had patients who had suffered years with asthma, hay fever and allergies. I had invalids and children with such weak constitutions their doctors couldn't find a way to help them. There were patients brought to me who were so close to the other side I wasn't sure whether I could help them. But I gave them fresh, raw milk, warm and foaming from the goat. What happened was almost unbelievable.

SEEING IS BELIEVING

You might have heard that "milk is milk," but it isn't true. I've had numerous patients who couldn't tolerate cow milk but only a handful that were unable to drink goat milk. Some people are so milk-logged that they can no longer tolerate any kind of milk such as goat milk or milk product such as cheese or ice cream. The majority would find goat milk easier to digest and of greater nutritional value than cow milk.

My sanitorium regimen for patients was centered around a balanced nutritional program with exercise, fresh air, sunshine, recreation and rest to relax, rebuild, refresh and renew those tired, sick people, who came to get over what they brought with them.

I was surprised at the rapid improvement of many of my patients when goat milk was added either as a recommended dietary supplement or an optional supplement. I had all patients with serious fatigue and tiredness using the fresh, warm goat milk. I had all my asthma and emphysema patients drink it. All the children had it and the underweight patients.

I feel there are several reasons why goat milk made these patients "come alive." First, goat milk has an abundance of minerals, electrolytes, trace elements and vitamins, which are needed to produce energy, make special proteins, form enzymes and do other tasks. It is easy to digest because its fat globules are much smaller than those in cow milk and because raw milk contains live enzymes that aid in its digestion. It has amino acids to repair damaged tissue, fatty acids for energy production and carbohydrates for energy conversion. Secondly, all the minerals and nutrients are at a highly-evolved level, ready to use in the human system. Third, the life force remains high in the fresh milk for three hours after it is taken from the goat, and when used as soon as possible, it strengthens the human system at all levels. Fourth, the brain is well-nourished by goat milk, which improves the brain's ability to direct all bodily functions and organs more efficiently. This speeds up healing. Lastly, my patients know that I believe in goat milk, and they drink it believing it is going to do them good.

IF IT WORKS, KEEP DOING IT

I had such wonderful results using goat milk with my patients at the Altadena Sanitorium that I continued to keep a small herd of milk goats when I moved to my Ranch near Escondido, California.

My patients aren't the only ones privileged to use goat milk. I have been taking it myself as a pick-me-up in midafternoon, sometimes mixed half-and-half with carrot juice. I seldom recommend a food to my patients or anyone else that I haven't used myself.

I recommend goat milk to all my students and patients, with the expectation that they will all benefit from it.

CHAPTER 3

THE HEALTH BUILDER

Wherever you go, you find goats. They may be in the minority, but there isn't a section of the country I have visited where I haven't found someone who has a goat, and the reason for having one is for health. Everyone we visit has either had a child who was very sick and needed that type of milk to get well, or they had had a child who at one time was very sick, and now they use goat milk because they find it is the best.

For young and old alike, it has proven to be the one food that is tolerated best by a disturbed digestive tract. The most sensitive tissues seem to welcome the administration of goat milk in the diet. The sensitive stomach can use the high sodium content of goat milk. A broken down liver can handle the naturally homogenized fat globules in goat milk. Since the fat cells are five times smaller in goat milk than in cow's milk, the liver finds it easier to handle this fat.

WHY GOAT MILK IS THE BEST

The high content of fluorine, which is the beauty element and the one element which gives the teeth their hard coating and protects the teeth from decaying, is found in raw goat milk. It is 10 times richer than in cow milk. Fluorine is a volatile gas which leaves the milk when too much heat has been applied.

The heat of pasteurization also destroys enzymes. For these reasons, I feel pasteurized milk is not as good as raw milk for human consumption.

We are getting a chemically-unbalanced milk when we don't get all the elements that the milk originally had. Stomach ailments respond in a short period of time with goat milk and carrot juice. The most sensitive stomachs and the weakest bodies thrive on this diet. Arthritis, rheumatism and any of our acid conditions are helped when the health of the body has been developed through the use of goat milk.

GOAT MILK, FRUIT AND WHEY

Soaked and peeled dates with goat milk is a wonderful combination for ulcers of the stomach. Black mission figs and goat milk seem to set well and help those with arthritis.

Goat milk is surely a vitality-producing food. Three-fifths of the world today lives on it and thrives on it. It is one of the great things to sweeten the intestinal tract and to get rid of constipation. To develop a sweet intestinal flora, use goat whey. This we can drink along with pineapple juice or any other sweet fruit juices.

At the Ranch, we have administered goat whey enemas, which helped many extreme gas cases.

Goat milk, next to mother's milk, is the best milk for feeding the baby. It compares in chemical balance to mother's milk. It has a chemical balance to feed baby goats which have more the structure of a human baby than does a calf. A calf has so much hide, hair and hooves to feed that the chemical balance in the milk is not as good for the human being as the chemical balance in goat milk. When man-made formulas do not work for the baby, try goat milk.

CLABBERED MILK AND YOGURT

Many of the people I met around the world with the greatest longevity and health used clabbered goat milk. The difference between clabbered milk and yogurt is that yogurt requires heating, while clabbered milk forms at room temperature and is still raw. Clabbered milk is more easily digested than regular milk while still retaining all the food value.

Milk should always be taken with other foods. If milk is taken alone, the gastric juice cannot penetrate the curd formed. Only 90% of the nutritional value is received. Never drink milk rapidly. Sip it slowly. Don't take it in large quantities or it will coagulate in the stomach like a lump of cheese. The high degree of acidity in the stomach favors the formation of tough curds. A person living on milk alone would have blue veins and lack magnetic qualities. About 8 pints of milk would be required to obtain enough iron each day. Goat milk is also short of folic acid and vitamin E, which must be obtained from other foods.

An acid-loving bacteria called lactobacillus acidophilus, capable of souring goat milk, is an excellent addition to the diet. When acidophilus is sufficient in the bowel, it not only sweetens the bowel, but reduces the quantity of toxic material by crowding out undesirable bacteria. The lactobacillus is capable of reducing the population of the undesirable bacteria E. coli.

Clabbered milk, which is raw milk that has been allowed to sour at room temperature, is an ideal food for the lactobacillus. Clabbered milk helps get the undesirable bacteria out of the bowel by feeding the desirable bacteria, which multiply rapidly enough to crowd the others out. Clabbered milk is easy to digest since its own acid content assists in protein digestion.

Dr. Charles Mayo has explained the difference between souring and putrefaction as follows: "Souring is a natural process in milk that comes with age, and to many is not objectionable. Souring prevents putrefaction. Preventing acid organisms which produce the souring makes the putrefaction process more rapid unless the milk is kept cold." He pointed out that putrefaction is a greater problem in pasteurized milk. Milk contaminated by putrefactive bacteria will cause sickness if it is consumed. Pasteurized milk lacks the lactobacillus, so it putrefies or spoils.

SEPARATING CREAM FROM WHOLE GOAT MILK

Cream can be separated from goat milk by using a simple law of nature. Add water to the milk, and the cream will rise to the top of the water and light milk. You can use a centrifugal separator like the dairy farmers do, but some nutritionists believe it is better to avoid the mechanical separation process. I believe it is better to use goat milk whole, since it is naturally homogenized and since nature designed our digestive systems to make better use of whole foods.

Goat cream is not as heavy or as yellow as cow cream. The carotene (pro-vitamin A) is higher evolved in the goat, and researchers tell us carotene helps prevent cancer. Goat milk contains both pro-vitamin A and complete vitamin A. In most body processes, vitamin A (not pro-vitamin A) is needed. The formation of vitamin A takes place as enzymes and bile salts change the pro-vitamin A into the complete vitamin A. This changeover is triggered by thyroid hormone. The thyroid of the goat is more highly evolved than that of most mammals. Vitamin A is absorbed twice as fast as pro-vitamin A, considering the time it takes for conversion into vitamin A.

Foremilk, the first part of the milking, contains more water and less fat. Strippings, the last part of the milking, have more fat and nearly 8 times as much cream. Albumin and milk sugar do not vary from foremilk to strippings. An overweight person should avoid the strippings. An underweight person should, by all means, take them if he can get them.

Loss of weight can be stopped by using goat milk and dried stewed fruits between meals. Using goat whey or skimmed goat milk will help those who want to reduce. Goat milk is alkaline in reaction because of its high sodium, potassium and calcium content. These chemical salts are so necessary in the human makeup. Because the goat is so fond of bark on trees and herbs and is really more of a browser than a grazer, we find the milk is high in silicon. This is one of the elements that is such an enemy of tuberculosis. We never find tuberculosis in goats or tuberculosis germs inn goat milk, while it is often found in cows.

WHAT THE OLD PEOPLE USED

All the old people I met during my world travels used "spoon milk," as they called it. This is a curdled milk. They made yogurt in Armenia, kumis in Turkey and matzoni in Russia. In all these countries, the whey from cheese-making is reserved as a drink for the older men. In fact, they call it "the old man's drink."

Older people digest their food more slowly. When milk goes into the stomach and meets the hydrochloric acid in the digestive juice, it curdles and becomes curds and whey. This happens normally in younger people. Older people, however, often lack hydrochloric acid. So, natural processes that curdle the milk before it is eaten make it more easily digested. This is why the older people use the "spoon milk" and get more good out of it than they would from fresh milk.

Whey, the liquid left from cheese-making, is high in food sodium. Food sodium is needed by the stomach walls and by the joints. Sodium protects the stomach walls from acid. It keeps the joints limber and flexible. They call sodium "the youth element." I have never found any old people with arthritis who regularly used spoon milk or whey regularly.

GOATS ARE BROWSERS, NOT GRAZERS

When I say goats are browsers, not grazers, I mean they are very particular about what they eat. I have seen a goat turn down a carrot after someone else had eaten a bite of it. I have seen goats turn their noses up at grass. It isn't true that goats will eat weeds, sagebrush and tin cans. They may nibble on the paper label of a tin can, just because they are curious or deficient in silicon, but they don't try to eat tin cans. A goat will sometimes stand on its hind legs in rich, green grass so it can reach the lower leaves on trees. They turn down the grass in favor of the leaves. They have an instinct for what is good for them. You practically have to starve a goat to get it to eat spoiled grain, food garbage or plant life that is deficient in minerals.

Cows, on the other hand, pick up unbelievable stuff when they are grazing. Baling wire, barbed wire, pieces of board, nails and broken glass have all been found in cow stomachs. Some farmers make their cattle swallow magnets so all the iron and steel they eat will collect in a safe place in their first stomach. Cows have four stomachs, as do goats.

A SOCIABLE ANIMAL

Goat milk is wonderful because it comes from an animal that makes a good pet. Goats love people and respond to affection. A goat needs the human touch more than a cow, and milk from a goat has a chemical balance like mother's milk. There is a nice rapport that develops between the goat and the milker. I feel this influences the chemical composition of the milk in a very beneficial way, and this may very well be one reason why goat milk is so good for human beings.

Many have found that goats can be treated as part of their own family, and they will respond. They are caring animals, sensitive and quick to recognize someone they know. You'll find out, too, if you put a stranger in with that goat, and the goat doesn't want him to milk her, she'll drop down to half the milk supply she will normally give—just through resentment...just through resistance. Goats have those mental attitudes just like people.

GOATS WERE HIS LIFE

The story of George Surdel, who lived on nothing but goat milk for more than 30 years, is an example of what goat milk can do if given an opportunity.

Goat milk saved his life in 1928, when he was near death in the mining town of Christopher, Illinois. He was seriously crushed by a rock fall when he was working in the New North Mine.

"The doctors patched me up, but I couldn't get my health back," Surdel said. "Then a friend told me about goat's milk. I tried it and my health miraculously returned."

After that, he decided to gather a few goats. He soon had more than 40 of the animals—bucks, does and kids. George took good care of his goats. He led them to a grazing area every day. Each night he rounded up the stray kids and returned them to their mothers.

For breakfast, Surdel would have a cup of warm goat's milk. At lunch time, he'd go out and milk another goat. In the evening, it was the same thing. He lived on nothing but milk from his goats.

Although Surdel had been a soldier in World War I before becoming a miner, he refused a veteran's pension. He also turned down Social Security. "I don't need anything from anybody," he said. "My goats take care of me in winter and summer. They're my life. I don't need anything else."

George Surdel had 42 goats and perfect health. They went together.

Goats are very inquisitive animals...or are they just plain nosey?

CHAPTER 4

GOATS ARE SODIUM ANIMALS

Goats are sodium-dominant animals, which is one of the reasons why goat milk is high in food sodium. This is the kind of sodium needed so much in the joints to keep them limber and in the gastrointestinal tract to neutralize acids and prevent indigestion. Goats are nimble creatures, sure-footed and quick. Wild mountain goats leap gracefully from rock to rock, never losing their footing. Domestic goats love to climb anything they can get on top of. I have a photo of a goat standing on the peak of the roof of the shed, looking very pleased and proud of herself. They have a wonderful sense of balance and coordination.

Each of the internal organs tends to favor one chemical element more than others in its activity. This element is stored more in the organ as a reserve. Dr. Koenig found sodium stored most in the stomach wall and stored second-most in the joints, then the lymph fluid. Sodium is needed to keep the stomach's digestive work at its best.

We find that very few people ever have enough food sodium in their diet.

SODIUM—"THE YOUTH ELEMENT"

Sodium is known as the "youth element." There are 345 milligrams of sodium in a quart of goat milk. Sodium keeps goats young, nimble and active, no matter how old they are. Animals need sodium to climb, jump, walk and run. Sodium in the bloodstream and lymph keeps calcium in solution. When sodium is short, calcium may begin to deposit in the joints. Lack of sodium in the bowel wall allows acids to irritate sensitive mucous membranes. We need enough food sodium in the diet to prevent sodium shortages in the body. Sodium shortages can even be found in children.

MY YOUNGEST ARTHRITIS PATIENT

They say even babies can be born with arthritis, but the youngest patient I ever took care of with arthritis was 10 years old. This little girl's mother carried her daughter in to see me. She told me her little girl had arthritis nearly all her life.

This 10-year-old girl was given Whex, a dried goat whey product high in food sodium. In about a year, the little girl was taking ballet lessons. The sodium in the goat whey (which is separated from the goat milk in cheese making), when taken into the joint liquid, allowed the calcium deposits to be dissolved back into solution in the lymph.

FOOD SODIUM IS NOT THE SAME AS TABLE SALT SODIUM

A chemist will tell you, "sodium is sodium," and I won't argue with that. But, I know that the sodium in foods is differently organized than the sodium in table salt, which can be a risk factor for heart disease and high blood pressure when too much is used. The average American uses 10-to-12 grams of table salt a day, which is totally unnecessary. We can get all the sodium we need from the food we eat.

Table salt is refined and purified at extremely high heat, leaving only sodium chloride in crystalline form. This is ground fine and mixed with a chemical additive to keep the refined salt from absorbing water from the air and clumping up in the salt shaker. Refined table salt acts more like a drug than a food.

Food sodium, on the other hand, is always found with potassium, calcium, magnesium and usually several trace elements. It is virtually never found alone in nature. Food sodium is in all fruits, especially the sun-ripened fruits (the sun is a sodium star), and in all vegetables, grains, legumes, nuts and seeds. It is found in dairy products and in meat, poultry and fish.

How is it that we find people who don't get enough food sodium if it is in all these things? It is possible that a person doesn't eat enough of the natural sodium-containing foods.

To avoid certain types of arthritis, digestive problems, irritable bowel syndrome and other sodium-deficiency problems, you may need to make fresh, raw goat milk a regular part of your food regimen. If your deficiency problem is long term, you may do well to use Whex, a more concentrated form of food sodium in a natural product containing many other minerals, electrolytes and trace elements.

I estimate that 30% of my arthritis patients get well if they stick to the food regimen I recommend to them. Another 30% show improvement, but there are 40% that don't seem to be helped.

THE SODIUM TYPE OF PERSON

The sodium type of person, which Dr. Rocine called a Desmogenic, is known for his or her youthful appearance into advancing age. Such a person, in good health, has limber joints and good digestion because he or she retains a good reserve of sodium, the "youth element," in the walls of the stomach and bowel, and in the joints.

A 160-pound man will have about 75 grams of sodium in his body. The liver contains about 0.02% sodium salts, the brain 0.2%, cartilage 0.9%, lymph and blood 0.7%, synovial membranes (in the joints) 0.8%, fibrous tissues 0.7% and the rest of the body's sodium is in the walls of the gastrointestinal system and in the bones. Bile salts carry several sodium compounds.

A "sodium type" is a person who metabolizes sodium faster than it can be replenished. Their complexion is pale, body lean, svelte, fragile appearing but with surprisingly tough ligaments, tendons and muscles. A sodium person is very active.

The sodium person is ambitious, hardworking, high-strung and chaotic. He is graceful, nimble, skilled and precise in his actions. A sodium person likes the outdoors, nature and the mountains, which suit him well. He has few friends and likes it that way.

A sodium person is always rushing, no matter where he is going or what he is doing. He is restless in sleep and easily angered in the morning.

Sodium people are prone to accidents, injuries, fevers, indigestion, fractures, sunstroke, pneumonia and other diseases and disturbances, due to disregard for temperate and health-supporting lifestyle principles. It is always difficult for the doctor to take care of a sodium person.

You may be a mental-motive person, a silicon person or a desmogenic type, like I am. I'm telling you, I move! If I had to walk up to my house tonight, I wouldn't walk all the way around. I'd jump the fence and go right over. That's my body and that's my chemistry. You find out you have to have the kinds of foods that feed that kind of body. I'm a sodium type. You may be a sodium type or something else. But, everyone needs the kinds of foods that fit their type.

When sodium deficiency is evident in a sodium person, they will respond well to a sodium-rich diet, if you can get them to follow the diet. It may be almost impossible to get them to drink 6-to-8 glasses of goat milk per day, but you may have better success getting them to take a couple of tablespoons of Whex, concentrated dried goat whey, dissolved in a glass of vegetable juice.

WHEY IS AN ANCIENT REMEDY

Hippocrates prescribed whey for an assortment of human ailments in the 5th century B.C. Hippocrates, called "the father of the healing art," also said, "If you want to be well, get a goat and live on the south side of the mountain."

Doctors have recommended whey to their patients for centuries. In 1761, Dr. Frederick Hoffman, physician to the King of Prussia, wrote, "There are few drinks more salutary and better adapted to prevent and cure diseases of the human body than whey." In the mid 1800s, there were over 400 so-called "whey houses" in Western Europe that offered their services in the cure of disease. Treatment consisted of having the patient drink up to a quart-and-a-half of whey per day, for ailments ranging from arthritis to liver disturbances.

I believe goat whey, either fresh or dried, is good for the bowels, lacteals, liver and kidneys, neutralizing acids and preventing the adverse conditions that develop in an over-acidic body from becoming a disease.

THE GOAT IS A FRIEND OF GOOD HEALTH

From Europe to the Mideast, from Africa to South America, the goat is considered a friend of good health, and goat milk drinkers are often known for their longevity. The food sodium in goat milk and goat whey is certainly a plus in favor of the health-protecting and disease-reversing capabilities of these products, but there is a whole array of other health-promoting nutrients in goat milk, as you will see in the next chapter.

CHAPTER 5

CATARRH—THE BEGINNING OF ALL DISEASE

Catarrh is the universal symptom of imbalances in the body, indicating disease-producing processes at work. It is always the first symptom to appear and it always indicates an excess of acids and mucus being developed due to tissue inflammation. Catarrh can be caused by a deficiency of the biochemical elements, an imbalance or excess of them or a toxic irritant in the body.

Catarrh is derived from the Greek words *cata* (down) and *rhein* (flow)—to flow down. It is the body's normal response to tissue inflammation and as long as it is allowed to flow, we know the body's natural defense system is working properly. Its presence often signals the onset of a cold, flu or other elimination process, but the most universal response to it is to run to the doctor or drugstore to get a drug to stop it.

We find that catarrh is not so much a cause of disease as a sign that something is wrong in the body. That is, it is part of the body's natural response to some other condition. Yet, when we suppress catarrh, when we drive it back into the body, we force it to collect in some organ or tissue where it becomes a host for germ life and viruses. Everything in the body must move to accomplish its function, and catarrh is no exception. When we stop the flow of catarrh by using drugs, we are planting the seeds of chronic disease.

Every "itis"—tonsillitis, bronchitis, sinusitis, appendicitis and the others—is fundamentally inflammation accompanied by catarrh. Healing cannot take place unless the catarrh formed to carry off toxic material is eliminated from the body. Then the inflammation will be taken care of by the body's natural immune system. Suppression of catarrh with drugs is not true healing, but only the temporary elimination of symptoms.

CATARRHAL CONDITIONS DEVELOPMENT

Diphtheria is a catarrhal disorder, characterized by development of a coating on the membrane of the throat that gives off a foul mucus.

Consumption is an advanced stage of chronic catarrh of the lungs, resulting from years of feeding on faulty food.

Catarrh often affects the Eustachian tube, leading from the throat to the tympanic membrane, or ear drum, causing deafness. It may result in suppurative inflammation or an abscess may form if not taken care of.

In women, the catarrhal condition may extend to the mucous membrane of the vagina, passing up into the uterus and out the Fallopian tubes, where the effect is in severe, delayed, painful or protracted menstruation. The ovaries may become involved, and the disorder is often treated with unnecessary surgery.

In both sexes, the catarrhal inflammation often extends to the kidneys and on to the bladder, causing bladder stones and the other disorders of the genitourinary system.

In fact, when the body becomes well saturated with the acid poisons resulting from poor food habits, a person will suffer from one or more of many catarrhal disorders named as follows:

- Catarrh of the stomach called gastritis.
- Catarrh of the mouth called stomatitis.
- Catarrh of the throat called diphtheritis (diphtheria).
- Catarrh of the nose called rhinitis.

- Catarrh of the bronchi called bronchitis (hay fever, asthma, etc.).
- Catarrh of the lungs called pulmonitis (influenza, pneumonia, consumption).
- Catarrh of the eyes called conjunctivitis (trachoma).
- Catarrh of the ears called otitis.
- Catarrh of the brain called meningitis.
- Catarrh of the small intestine called enteritis.
- Catarrh of the large intestine called colitis.
- Catarrh of the appendix called appendicitis.
- Catarrh of the liver called hepatitis.
- Catarrh of the pancreas called pancreatitis.
- Catarrh of the kidneys called nephritis (Bright's disease).
- Catarrh of the vagina called vaginitis (leucorrhea).
- Catarrh of the uterus called metritis.
- Catarrh of the ovaries called ovaritis.
- Catarrh of the bladder called cystitis.
- Catarrh of the prostate called prostatitis.
- Catarrh of the joints called arthritis.
- Catarrh of the veins called phlebitis.
- Catarrh of the arteries called arteritis.
- Catarrh of the heart called carditis, pericarditis, endocarditis, etc.

The above list, while far from complete, serves well to show two things: (1) how medical institutions name immaterial symptoms according to their location and foolishly treat each symptom as a different disease while disregarding the cause responsible for the symptoms, thus leaving the cause to continue its degenerative work; (2) how one harmful habit, although apparently insignificant, when steadily pursued deranges the whole body, establishing slowly but surely a chronic condition that finally affects every organ, structure and function.

As the body normally functions, mucus is made by goblet cells in the mucous membranes lining the alimentary system, the respiratory system and parts of the genitourinary system. This mucus lubricates and protects the sensitive tissue lining these parts of the anatomy. If germ life, foreign matter or toxic substances enter the body, they are entrapped in the sticky mucus which is eventually excreted from the body as flowing mucus.

While I have listed the different "itises" that doctors are concerned with, I have found that cow milk is contraindicated with any of these conditions because it aggravates the inflammation. But I can give goat milk to the patient with any of these conditions, and they will get well. For instance, bronchitis can often be eliminated by changing from cow milk to goat milk.

There may be a few cases in which goat milk can't be used by particular individuals. However, goat milk is similar human needs then cow's milk. When goat milk is used, we don't have to worry about catarrhal troubles.

While a lot of children are plagued by children's diseases, it is often because they are using too much milk, wheat and sugar in their diets. When I encounter an "itis" in a patient, I immediately cut out these three foods and recommend substitutes. Milk, wheat and sugar are catarrh producers because they are overused and body tissues are irritated by them.

Mucus is formed in response to irritation, and unless it is eliminated, it becomes a problem. Mucus is a sticky semi-liquid protein which traps allergens, toxic substances and germ life so they can be ingested by white blood cells and eliminated from the body. However, when a suppressant drug is taken, the mucus dries out and the toxic material it carried is absorbed back into the tissue, preparing the way for an even more serious condition to develop later. Catarrh should never be treated by suppression.

STOP EATING CATARRHAL-PRODUCING FOODS

Junk food diets are among the most common causes of catarrh. There is so much material in junk food that the body can't digest or assimilate that it must be eliminated as waste. Diets high in certain starches and proteins can produce catarrh.

Over 50% of the average American diet consists of wheat and milk products, and any allergy specialist can testify that these are heavy catarrh producers in many people. On a visit to China at one time, I noticed people spitting everywhere, probably due to a concentration of white rice in the diet. By and large, Oriental people avoid the more nutritious brown rice, calling it "dirty rice." I believe many of them would rather starve than eat brown rice, but the price they pay for their heavy white rice diet is catarrh.

Breathing polluted air can cause catarrhal conditions in the lungs. Any external irritant taken into the body will produce catarrh.

I have found some persons who produce catarrh much more easily than others, no doubt due to a sensitive metabolism or perhaps a lack of certain enzymes. In fact, it is possible for the body to load up with acidic wastes and catarrh simply because of one or more underactive elimination channels.

Normally, the body is busily replacing old cells with new ones, and metabolic wastes are carried away at roughly the same rate that new nutrients and building materials come in. But, whenever one of the elimination channels is underactive—bowel, kidneys, lungs, skin or lymphatic system—the others must try to pick up the overload or else catarrh builds up in the body. The bowel is the most frequent problem, but any of the elimination systems can be involved. Without sufficient exercise, the lymph doesn't move fast enough, and accumulations of waste and toxic materials can build up at the lymph nodes in the groin, under the arms, in the neck and elsewhere. Kidney problems are not uncommon, particularly since they do often develop as side effects from conditions like atherosclerosis and diabetes. It is possible for cholesterol to block the fine blood capillaries under the skin, reducing the elimination via the skin. There are many ways in which the efficiency of the eliminative organs can be reduced. Smoking and polluted air, of course, hinder the lungs from expelling toxins and catarrh. If any of the elimination systems are slowed down, catarrhal buildup in the body can be due to metabolic wastes alone, even if the diet is basically good.

Genetic factors may contribute to catarrhal conditions. Everyone is born with certain strengths and weaknesses in the body, but the strengths don't create problems, so we only have to take care of the weaknesses. When the digestion or assimilation are weak, more catarrh is produced. When any of the eliminative organs is underactive, more catarrh is produced. When certain enzymes are lacking or when the thyroid is underactive, more catarrh is produced. Chill or fatigue affects the weaker organs and tissues more severely than other parts of the body, allowing catarrh to develop.

In my experience, one of the most common causes of catarrh production is fatigue. A tired body is not eliminating well, and it develops acids that are not quickly removed from the body.

We find that catarrh does not become a problem in the body as long as we keep it moving. In terms of physiology, the life of the body is in the constant movement of the thousands of processes that keep everything going. Movement is life; stasis is death. When anything slows down in the body, irritation develops and catarrh is produced. If catarrh itself slows down or stops, it irritates tissue and promotes more catarrh.

We find also that slowing down any process in the body is affected to some degree, by a single change. If nutrients are stopped, cells starve, die and begin to decompose almost immediately, resulting in more catarrh. Of course, bacteria and germ life thrive on dead matter, and more catarrh is produced as the body's natural immune system goes to work against them.

It would be safe to say there are a thousand and one causes of catarrh. But there is one fundamental and appropriate response to catarrh: let it run—or better yet, help the body get rid of it faster.

CONSEQUENCE OF CATARRH SUPPRESSION

Over the past few decades, the elimination of the common cold and other catarrhal conditions have almost become a crusade with the drug manufacturers. Drug companies, complete with their TV ads over which drug gives the fastest, most effective "relief" of symptoms. By relief, they mean suppression.

The common cold is regarded as a viral disease by Western medicine, but doctors cannot account for why some people are so susceptible to colds, while others never get them at all. It apparently doesn't occur to them that diet, constitutional strength and inherent weaknesses determine the degree of tissue response to any irritant—and whether or not inflammation and catarrh will develop. A virus does not "cause" anything. We are constantly exposed to viruses and bacteria of many dangerous types, yet we seldom are affected by them unless our bodies become depleted by nutrient deficiency and tissue becomes inflamed. Germ life must have substance to feed on or it cannot thrive and reproduce.

According to experts, the viruses differ greatly among individuals with colds, which is why a vaccine can't be developed for the common cold. As we grow older, our immunity to colds appears to increase. We should not take suppressant drugs for cold symptoms but let the catarrh which develops flow out of the body.

To suppress catarrh requires driving it back into the tissues of the body, along with all the germ life, foreign matter and toxic accumulations it contains. There it functions as a low-grade poison, interfering with tissue metabolism and serving as a continuing irritant. The usual consequence is the development of chronic disease, the type of disease depending on the inherent weaknesses of the individual.

Most Americans fail to realize the price they pay for suppressing catarrh. A cough, for example, is a natural reflex action to rid the upper bronchials of catarrh. From the advertisements of some cough medicines, we find they work by suppressing the cough center in the medulla of the brain, where there are also centers relating to the heart, lungs and digestion. How can a drug affect the cough center without affecting the other vital centers in the same brain area? Other drugs act to dry up catarrh in the body, inviting the development of abnormal tissue pathology wherever the dried catarrh settles.

We find that the first operation most people have is a tonsillectomy. The tonsils are lymph tissue, and tonsillitis is an inflammation of the lymph tissue due to toxic overload. To remove the tonsils is like trying to solve a garbage problem by getting rid of the garbage trucks, and it doesn't work. After the tonsils, the next operation is often an appendectomy, and the appendix is also lymph tissue. Each time lymph tissue is removed, the body has to find some other place to send catarrh and lymphatic waste. So, we encounter fibroid tumors and cysts in the breast tissue (also lymphatic) or in the lungs, stomach, bowel or pelvic area. The catarrh **must** go somewhere when it is suppressed, and it always causes problems when it is not allowed to flow out of the body.

Suppression of catarrh is always an invitation to disease.

THE FOUR STAGES OF DISEASE

All diseases go through four stages: acute, subacute, chronic and degenerative. The **acute** stage is an active stage of disease, usually accompanied by catarrh, fever, coughing, inflammation and soreness localized in the body. Catarrh may issue from any orifice of the body, and strands of it may be found in the urine. In the acute stage, the body is actively trying to throw off the disease, and all organs are hyperactive to support the elimination.

The **subacute** stage is more serious. If the body fails to throw off the disease in the acute, running stage, the inflammation sinks deeper into the tissues, reducing the metabolic rate and further weakening the body.

The **chronic** stage finds the body in a lower condition yet. This is where flu, colds and hay fever turn into asthma, with occasional bouts of pneumonia. This is where joint aches turn into chronic arthritis.

The final or **degenerative** stage is very nearly the point of no return. Asthma turns to emphysema. The joints swell and grow, calcium spurs appear, in the arthritis case. Lumps and tumors are found to be malignant.

Most poor health in adults can be traced back to childhood conditions. All children's problems such as colds, flu, earaches, mumps, measles and tonsillitis should be taken care of by natural means to prevent catarrh from being driven back into the body. Cleanliness, correct foods, stimulating activities and pleasant surroundings, with plenty of love, will keep most children well. Herbal and other natural supplements may be needed from time to time, but suppressants must be avoided.

Nature's way is to let catarrh run until the source of irritation and inflammation is thrown off.

ELIMINATION OF CATARRH

Hering's law of cure says, *All cure starts from the head down, from the inside out and in reverse order as symptoms have first appeared.* Each time we have built the body up through right nutrition, exercise, recreation and rest, to the place where it reaches a certain level of strength, a healing crisis occurs in which old symptoms of disease are retraced and catarrh is liquefied and thrown off. Depending on how chronic or degenerative the disease has become, it may take several healing crises to eliminate the old catarrhal deposits.

I have never had a person under my care for a year or more who did not notice better bowel function, fewer colds and higher well-being than in many years. Any who have experienced catarrhal discharges from any part of their body notice definite improvement. Properly selected foods make a decisive difference in ridding the body of catarrh.

Lord Chichester, after discovering he had cancer, went on a grape diet and sailed around the world. By the end of his trip, the cancer had disappeared. Asthma can be taken care of it we can get rid of the miasmic background present in asthmatics. Some of this may even have been inherited from their parents. Likewise, to get rid of colitis, an inflammation of the large bowel, we must get rid of the mucus strings in the colon. Much of the process of healing, no matter what the disease, requires elimination of catarrh from the body.

AN OUNCE OF PREVENTION

It is time we realize there is a way of living which will bring the cell structure of the body to the highest possible integrity. The great majority of people walk around living between a subacute and chronic condition, unable to do their work competently, be on time, think efficiently, remember well or enjoy their lives. Low-grade infections are tying up half their energy, and their state of well-being is flying at half-mast. It is all so unnecessary.

When we eat right, exercise properly, get enough sunshine, fresh air and rest, think positively and live so we can lift others into a better way of life, our own lives reach the state of high-level well-being that makes it a joy to be alive.

CHAPTER 6

THE SUPER NUTRITION OF GOAT MILK

In this day, when science and the government have finally discovered there are foods that heal and foods that prevent disease, we can expect goat milk to become a subject of research for its particular value to healing and health.

We know that goat milk is a wonderful healing aid. What we don't know is exactly how it does what it does. We have found out, by trial and error, that goat milk assists in the healing of certain diseases, such as ulcers, tuberculosis, arthritis, rickets, irritable bowel syndrome, kwashiorkor, nervousness, anorexia, asthma, hay fever, allergies and other conditions. But, because research on the healing effects of goat milk is lacking, we don't really know how it produces such wonderful results.

A SURGEON'S TESTIMONY

Dr. William Lee Secor, a medical doctor from Kerrville, Texas, wrote:

> "I am a surgeon, restricting my practice to surgery, but since discovering the value of goat milk in the treatment of stomach and intestinal ulcers, I have not operated on a single ulcer case, and this covers a period of a dozen years."

We need to recognize that milk is a whole, pure and natural food (when taken from a healthy animal), and it contains all the nutrients necessary to nourish a newborn animal the first year of its life.

It is one of several foods that are able to provide all the vitamins, minerals, enzymes, protein, carbohydrates, fats and other nutrients to sustain life.

Other such foods are eggs, fish roe, seeds and honeybee pollen. They are able to build a whole life, and often they are found to have healing properties and to provide glandular support. Dr. V.G. Rocine believed that fresh, raw goat milk was especially supportive of the sexual system.

A quart of goat milk provides 630 calories per day, which would meet 24% of the average man's daily energy needs and 39% of the average woman's daily energy needs.

The percent of energy needs met for the average teenage boy would be 21% and for the average teenage girl, 27%. We find that teenagers have a higher metabolism than adults, and that the energy required for growth and for sex-related endocrine changes use up more energy than is the case with older adults.

ENERGY, GROWTH AND TISSUE REPAIR

From the protein, carbohydrates and fat in our foods, we get the energy and materials for growth and tissue repair. These are the substances that account for most of the carbon, nitrogen, sulfur, hydrogen and oxygen in the body, except for water, which is two parts hydrogen to one part oxygen. About 70% of the body is water, and the second-most plentiful substance is protein.

Protein is needed to form the lace-like matrix that calcium compounds are deposited on to form bones. We need protein to form muscle tissue, blood, brain tissue, skin, hair and nails. Most vital organs have muscle tissue made up of protein. Substances like hormones, enzymes, antibodies and

NUTRIENTS IN ONE QUART OF GOAT MILK (and the Percentage of Daily Adult Requirement)

Nutrient	Amount	RDA Daily (%)
Protein*	30 gm	54.0
Carbohydrate*	43 gm	11.0
Fat*	38 gm	44.0
Calcium	1213 mg	100.0
Phosphorus	996 mg	125.0
Sodium**	345 mg	7.0
Potassium**	1794 mg	90.0
Iron	1.0 mg	10.0
Zinc	2.3 mg	7.0
Chloride**	1128 mg	38.0
Magnesium	115 mg	33.0
Copper	0.1 mg	5.0
Manganese	0.02 mg	0.5
Vitamins: A	1950 I.U.	39.0
B-1	0.4 mg	33.0
B-2	0.17 mg	10.0
B-3 (niacin)	1.8 mg	14.0
B-6	0.7 mg	35.0
B-12	0.6 mcg	12.0
Pantothenic acid	3.2 mg	32.0
Folic acid	5.6 mcg	1.4
Biotin	37 mcg	12.0
Choline	141 mg	28.0
Inositol	197 mg	39.0
C	14 mg	23.0
D	23 I.U.	6.0
E	None	0.0

*The protein, carbohydrate and fat requirements were calculated for a 154-pound man at light activity. The corresponding figures for a 128-pound woman at light activity would be: Protein 65%, carbohydrate 14% and fat 58%.

**National Research Council has not established any RDA for these. Percentage figure is based on percent of minimum average daily intake.

immunoglobulins are made of protein. Protein, carbohydrates and fat can all be transformed into energy. All protein is made up of smaller building blocks called amino acids. Four of the amino acids—methionine, taurine, cystine and cysteine—contain most of the sulfur in the body. Protein is found mostly in meat, poultry, fish, milk and eggs, and to a lesser extent in nuts, seeds, grains, legumes and vegetables.

Carbohydrates are the biggest source of energy for our body functions and physical activities. They help control protein and fat metabolism. The liver can't break down fats without carbohydrates. Heat, a form of energy, is also chiefly produced from carbohydrates. The basic carbohydrates are starches, sugars and cellulose, and they are found in most foods. Cellulose is not broken down in the human digestive tract but serves as a useful bulk in maintaining a normal bowel transit time and a clean bowel. All starches and sugars are eventually broken down to glucose, sometimes called "blood sugar." This glucose is the main form of energy fuel for the brain, nerves and muscles. Stored fatty tissue can be broken down, transformed into glucose and used to provide energy.

Fats are twice as concentrated in energy per gram as carbohydrates and proteins. Fats release 9 calories per gram when converted to energy. Protein and carbohydrates release 4. Fats act as carriers for fat soluble vitamins like A, D, E and K. The change of carotene into vitamin A would be impossible without fats. Fatty tissues insulate and hold the vital organs in place. Cholesterol is a fat-related substance that is needed to form vitamin D, hormones, bile, protective coatings for nerves and other substances in the body. Vegetable and seed-derived oils which remain liquid at room temperature are called unsaturated fats or oils. Saturated fats are mostly derived from animals and these become solid at room temperature. The National Research Council recommends that the amount of fat in the diet be restricted to 30% of the daily calorie intake.

Goat milk protein, carbohydrates and fats are all easily digested. The fat particles are 1/5 the size of fat particles in cow milk. The protein is of a superior kind, and the carbohydrates are sugars that quickly convert into energy.

MINERALS IN GOAT MILK

The most plentiful mineral ingredients of goat milk also happen to be electrolytes, which are very important to the health and well-being of our bodies. These are sodium, potassium, magnesium, calcium, chloride and phosphorus. In the body, electrolytes make up the electrically-charged ions that help regulate water balance, acid-alkaline balance, osmotic pressure, nerve impulse conduction, muscle contraction and transport into and out of cells. Most of the potassium, magnesium and phosphate are inside the cells, while most of the sodium, calcium and chlorine are in the lymph or interstitial fluid outside the cells. When electrolyte levels drop below a certain level, body functions are seriously disturbed. This can happen due to prolonged vomiting or diarrhea, which is more dangerous for infants than older children or adults. Infants can die from depleted electrolytes.

The minerals in goat milk include calcium, phosphorus, sodium, potassium, chloride and magnesium. Let's look at what each of them does in the body, then move on to the trace elements. Trace elements are technically minerals, but they are called trace elements because there is just "a trace," a tiny amount of them, in the entire body.

We all know that vitamins are important but we need to realize that vitamins and minerals work together to build up and benefit the body. Without minerals, vitamins are useless.

Calcium, "the Knitter." This element in foods requires sufficient hydrochloric acid in the stomach to be dissolved and prepared for absorption in the small intestine. Usually 20-30% of the calcium taken in foods is absorbed. One quart of goat milk has about 1,200 milligrams of calcium, the amount needed by teenage boys and girls, and more than the amount required by adult men and women. Calcium absorption depends on the presence of vitamins A, C and D, and can be blocked by oxalic acid foods (chocolate, chard, rhubarb) and phytic acid foods (cereals and grains). Calcium is mainly used in the bones and teeth, together with phosphorus, magnesium, sodium and many

One of nature's heroines.

Dr. Jensen with some of his goats in the present Ranch herd.

Hidden Valley Health Ranch goat herd from a number of years ago.

Dr. Jensen picking figs on his Ranch. Figs are one of the highest sodium-containing foods, along with goat milk, to help prevent arthritis. Fresh, raw figs with goat milk is a meal itself.

Goats at Dr. Jensen's Ranch. Kidding time is always an exciting time each March and April.

Goats on the Ranch have an ideal habitat. There are rocks to trim hooves and oak trees to nibble the leaves and have the best shade from the hot sun.

Dr. Jensen with new kids on the Ranch, along with Apolinar, the man in charge of caring for and milking of Dr. Jensen's goat herd.

This llama was put in with the goats and it took over as the protector. In fact, a mountain lion came in one night and when the owner went to get his gun he returned to find that the llama had killed the mountain lion.

Goats from the dairy where Dr. Jensen got the milk that saved his life recently.

Goats are very popular around the world. The top picture was taken in Switzerland, where the caretaker moved his goats through the town to browse in the mountains. The bottom picture is a herd of goats in New Zealand.

Notice the difference in the ears of the French Alpine goat in the top picture, and the Nubian in the bottom picture. Wherever you go, you can find goats. They may be in the minority, but there isn't a section of the country I've visited where I haven't found someone who has a goat, and the reason for having it is for health.

other minerals. I call calcium "the knitter" because of its function in healing. A small amount of fluoride is needed to harden the outer surfaces of the teeth and bones. Calcium helps form blood clots, neutralize acids, regulate the heart rate, relax the nerves, contract the muscles, activate enzymes and helps the body use iron properly. Goat milk is the best source of calcium I know, and 1,200 milligrams is a generous amount. Because of the current epidemic of osteoporosis among older women, the ladies especially need to make sure they are getting plenty of calcium. Tooth decay may show a lack of calcium or fluoride or both.

Phosphorus, "the Light Bearer." Nearly 1,000 milligrams of phosphorus is in a quart of goat milk, which is good, since calcium needs an equal amount of phosphorus to be properly absorbed. The phosphorus in goat milk is highly evolved, possibly the best for use in the brain in its control of the human body. The nickname for phosphorus is "light bearer." The brain needs the highest-evolved phosphorus, which is found in goat milk. The bones can use the lower-grade phosphorus from seeds, nuts, grains, vegetables and fruits. Phosphorus is seldom found deficient in the diet or human body. Phosphates are essential for the production of energy in the body and in the use of protein, carbohydrates and fats for physical growth and repair of damaged tissues. Phosphorus stimulates the heartbeat and is needed in DNA and RNA. Lecithin needs phosphorus to help balance cholesterol and to transport fats in the bloodstream. All proteins are high in phosphorus.

Sodium, "the Youth Element." The kind of food sodium our bodies need is best supplied in goat milk and goat whey. I've seen miracles done in the removal of arthritis spurs from the neck and spine by patients who faithfully used their goat milk or goat whey every day. Sodium in the bloodstream helps keep calcium in solution so it doesn't cause trouble in the joints. Sodium is called the "youth element." Would you be surprised to know that 50% of the body's sodium is in the bones? The rest is in the lymph, blood and the fluid that surrounds the cells. Sodium works with potassium in the transmission of nerve impulses, the regulation of the acid-alkaline balance, the osmotic pressure of cell membranes and in muscle contraction. Sodium is stored in the walls of the stomach and bowel and in the joints. When we have digestive problems or joint troubles, we usually find out there is a shortage of blood sodium. (Table salt sodium is more a drug than a food, in my opinion, and should not be used. Try herbal and vegetable seasonings instead.) Excess sodium drives potassium out of the body. Goat milk has 345 milligrams of sodium. The National Research Council recommends a daily intake of one gram per kilogram (about a quart) of water consumed. I say this should be food sodium—goat milk is a good source, so is goat whey, green vegetables, okra, celery and strawberries.

Potassium. This element I call "the great alkalizer," because of its work in the heart and all other muscles of the body, neutralizing acid wastes and keeping the acid-alkaline balance constant, together with sodium. Potassium neutralizes acids in the muscle tissue, including the heart. A heart can be kept alive 15 minutes by running sodium solution through it. If the same thing is done with potassium, the heart stays alive for 3-1/3 hours. The heart is a potassium organ. All leafy, green vegetables are high in potassium and so is goat milk. The amount of potassium in a quart of goat milk is 90% of the body's need for the day. Potassium does many of the same things sodium does, but it also unites with phosphorus to carry oxygen to the brain. (Vitamin E also seems to increase the oxygen to the brain.) Potassium is needed for skin health, for converting blood sugar to a form that can be stored in the liver. It is said to stimulate the kidneys to eliminate toxic matter from the body. The 1,794 milligrams of potassium in a quart of goat milk is the highest quality of food potassium you can find.

Chlorine. "The cleanser" is the name by which this element is known. Chlorine is often found in nature together with sodium or potassium or both. It combines with them neatly, since it has a single negative charge and both sodium and potassium have a single positive charge. Together with hydrogen, chlorine forms hydrochloric acid which is necessary to digest proteins and to prepare calcium for assimilation. It also helps regulate the acid-alkaline balance. There is about 1,100 milligrams of chloride in a quart of goat milk, 38% of the usual daily intake from all sources. When chloride is low in the body, there is loss of appetite, mental apathy and cramping of the muscles. This is seldom found, since chloride is abundant in foods.

Magnesium. I call magnesium "the relaxer" because it helps relax neuromuscular contractions, including contractions that cause the heartbeat. Calcium causes the contraction, magnesium releases the relaxation of the muscles. Magnesium neutralizes acids in the bowel and unites with calcium, phosphorus and other elements to form the bones. Inside each cell in the body, magnesium activates the enzymes required to metabolize amino acids and carbohydrates. (There are over 3,000 enzymes in every cell of the body.) Magnesium promotes absorption of calcium, phosphorus, sodium and potassium. Conversion of blood sugar to energy and regulation of body temperature are both assisted by magnesium. The amount of magnesium in one quart of goat milk is 115 milligrams, enough to provide 1/3 of the body's daily need. The average U.S. diet reveals a borderline condition in which the daily need of 350 milligrams is barely met. Symptoms of deficiency include muscle twitching, tremors, anxiety and confusion. The danger of magnesium deficiency is increased in all who use fortified, pasteurized milk with artificial vitamin D in it. This type of vitamin D, called calciferol, binds to magnesium and takes it out of the body. (If you need a vitamin D supplement, take fish liver oil, not synthetic D.)

This concludes our discussion of the minerals in goat milk. The only minerals in this group that we may fall short in are the food sodium and the magnesium. Later, I'll introduce you to a wonderful food sodium source named Whex, which happens to be made of dried goat whey. But, right now our first trace element is going to be iron, and you'll soon find out that the way I boost the iron content of goat milk also boosts the magnesium content.

TRACE ELEMENTS FOR BIG PROTECTION

The trace elements most often included in analyses of goat milk are iron, zinc, copper and manganese, but we have recently found out there are others of significance, including boron, chromium, nickel, sulfur, selenium, silicon, tin, vanadium, iodine and fluorine. I don't have quantitative amounts of the latter elements, but I can tell you what they do when present in the body.

Iron. I call iron the "frisky horse" element because it works with oxygen to oxygenate the blood and help energize the body. Goat milk has 1 milligram of iron per quart, which makes only 10% of the RDA. To make up for this lack, I add one or two tablespoons of blackstrap molasses to a glass of goat milk. Each tablespoon of blackstrap molasses adds 3.2 milligrams of iron (and 52 milligrams of magnesium) to the milk. You can very well use both minerals.

Manganese. A quart of goat milk has 0.02 milligrams of manganese, which is only half of one percent of what the average person takes in per day. There is no official RDA for manganese, but I call it "the love element." Mother rats deprived of manganese rejected their offspring in a study done some years ago. Manganese is needed to activate enzymes, synthesize fatty acids and cholesterol and stimulate normal development of the bone structure. It takes part in sex hormone production, nourishes the brain, aids in producing mother's milk and is involved in the metabolism of protein, carbohydrates and fat. Good food sources are eggs, green vegetables and whole grain cereals. The best source is Missouri black walnuts.

Iodine. Iodine, "the metabolizer," is needed in the amount of one microgram per day, or one part per million. There is a trace in goat milk, but it is well to take a little Nova Scotia dulse, and eat seafood now and then to make sure you are getting enough. Those lacking in iodine are angry people, mean and irritable. Sometimes they get claustrophobia just sitting in a room with the doors and windows closed. The hair, nails and skin are dry. There is iodine in most commercial salt, but since

I recommend not using table salt, you should get your iodine from foods. It is needed exclusively by the thyroid for the formation of thyroxin, a hormone that is so important that deficiency should be avoided at all costs. Deficiency can lead to a host of adverse symptoms including goiter, weight gain, obesity, fatigue, sluggishness, heart palpitations, nervousness, constipation, hardening of the arteries and many others.

Zinc. The RDA on zinc is 10 milligrams and a quart of goat milk has 2.3 milligrams, 23% of the daily allowance. That's not bad, as long as you get more zinc from your other food sources. Deficiency may cause birth defects, slow growth and delayed healing. Because zinc deficiency is common in people over 40, make sure you are getting enough. Zinc is used in the prostate, bones, pancreas, liver, kidneys, spleen, retina and skin. Zinc is needed by 70 enzymes. It is used to make DNA, RNA and protein. It is a component of insulin and of male spermatozoa. Zinc helps metabolize protein, carbohydrates and fat. Zinc is found in the white blood cells of the immune system. This is an element you don't want to put yourself at risk over. Organ meats are high in zinc, as are pumpkin seeds, wheat germ and whole grain cereals.

Copper. We need 2 mg of copper daily, but goat milk supplies only 5% of that in a quart. Copper is needed to form hemoglobin, metabolize protein, assist in healing, produce RNA and make the nerve sheaths that protect most nerves. It is used in over 12 enzymes. Best copper foods include liver, legumes, seafood, whole grains and green, leafy vegetables.

Sulfur. The National Research Council says we get all the sulfur we need from our proteins. The sulfur-containing amino acids in protein are taurine, cysteine, cystine and methionine. Since a quart of goat milk has 30 grams of protein, I suppose we get quite a bit of it. Sulfur is found in the skin and hair, in insulin and in muscle tissue. It takes part in tissue respiration, collagen synthesis and plays a part in the secretion of bile. Rocine taught that sulfur cools the brain and is needed for general well-being. Good food sources include meat, fish, legumes, nuts, eggs, cabbage, Brussels sprouts and cauliflower. Fair-haired people have a good deal of sulfur in their bodies.

Selenium. This trace element is needed in the amount of 70 micrograms daily. That's 70 parts per million. Selenium works with vitamin E to destroy cancer-producing free radicals in the body. Studies have shown it reduces the risk of cancers of the colon, rectum, breast, ovary, prostate, pancreas, bladder, skin and lungs. It also reduces risk of leukemia. Heart attack and stroke risk are reduced in those who take selenium. The protection of vitamin E and selenium includes guarding cell membranes from damage and increasing immune system activity. It also detoxifies mercury, cadmium and peroxidized fats. You can't trust foods to give you enough selenium unless you know for sure that the soil they are grown in has enough selenium. You should consider taking selenium supplements from your local health food store.

Silicon. This element is common in the soil, common in our foods, and we get about a gram a day. There is no RDA on it. Soluble fiber like pectin, containing a lot of silicon, reduces blood triglycerides and cholesterol. It is found in all collagen and cartilage, in the bones, hair, skin and nails. It may play a role in healing. Silicon foods, beside goat milk, include rice bran syrup, oat straw tea, whole grains, asparagus, beans, parsnips, strawberries and watermelon. Eat plenty of fruits and you'll get plenty of silicon. When I buy a goat, I feel its hair. It must have a nice coat of hair because then I know it has plenty of silicon in the body. That means there will be silicon in the milk.

Boron. Whether boron is recognized as an essential trace element or not, it is found in the human body, and it is believed to enhance mental functioning. There is a little in goat milk.

Fluorine. This element helps harden bones and teeth and acts as a natural germicide in the body. I call it the "resistance element" because it helps us resist colds, flu and infections. If you don't have enough fluorine, you can't resist diseases caused by bacteria. There is a little in all goat milk. There is more in Florida black bass and in green quince, but who wants to eat them raw? Cooking destroys the fluorine. The RDA is not established, but one microgram per day is considered sufficient. It is found in many foods, especially seafood.

Chromium. There is no RDA for chromium but its deficiency is so common that we should make a considerable effort to get at least the 70 micrograms daily that the body needs. Chromium activates enzymes that improve the effectiveness of insulin. We need chromium to utilize our blood sugar more effectively.

Nickel. In the past few years, experiments have shown that animals need a trace of nickel to avoid an array of deficiency symptoms. There is a little nickel in goat milk, and we also get plenty from a normal diet.

Vanadium. A little of this element is in goat milk. It seems to prevent cholesterol formation in the blood vessels of the brain. Ocean fish and seafood are the best sources.

Trace minerals are now known to be extremely important for the prevention of disease and to promote the best level of health we are individually capable of reaching. The best way to get all the trace elements we need is to use goat milk, seafood and emphasize variety in our daily food program. Use as many raw vegetables, nuts, seeds, fruits and sprouts as you can conveniently tolerate in your diet. You get the most nutrition out of raw foods because heating or cooking destroys some of the minerals, trace elements and vitamins. We will now explore the vitamin values in goat milk.

VITAMIN POWER IN GOAT MILK

In our chart at the beginning of this chapter, I show there are generous amounts of vitamins in a quart of goat milk. Two vitamins, however, are in short supply or missing. The amount of folic acid is only 1.4% of the RDA, and there is no vitamin E. So we have to get these from other foods or supplements. So, let's find out what vitamins are in goat milk and what they do.

Vitamin A. In one quart of goat milk, we find 1,950 I.U. of this vitamin in a combination of vitamin A and carotene (sometimes called pro-vitamin A). This is nearly 40% of the daily requirement. Pro-vitamin A is anti-carcinogenic. Vitamin

A is needed by the eyes to prevent night blindness, the growth and repair of tissues in the body, protection of the mucous membranes of the nose, throat and lungs and is used to trigger the secretion of gastric juices for digesting proteins. Vitamin A helps build strong bones and teeth and is used in the synthesis of RNA in each cell nucleus. Norwegian cod liver oil is the best supplement if you need additional vitamin A in your diet.

Vitamin B-1 (thiamine). B-1 helps convert carbohydrates to blood sugar, improves digestion and assimilation and supports the brain functions that have to do with mental attitude and learning ability. Rice bran syrup, blackstrap molasses and wheat germ are good sources. A quart of goat milk supplies 33% of the RDA of this vitamin.

Vitamin B-2 (riboflavin). Ten percent of the RDA of this vitamin is supplied by a quart of goat milk. B-2 works with a group of enzymes to break down and use protein, carbohydrates and fat. It supports cell respiration and is needed by the skin, nails and hair. Another good source is organ meats.

Vitamin B-3 (niacin). A quart of goat milk provides 14% of the average adult's need for niacin. B-3 acts to support the nervous system, the synthesis of sex hormones, reduce cholesterol production in the liver, and stimulate circulation of the blood in the capillaries. Taking niacin may result in a flushing of the face and upper body with a sensation of tingling that some find uncomfortable. However, it is harmless. Niacin is also needed for a healthy skin, tongue and digestive system.

Vitamin B-6 (pyridoxamine). This vitamin is a natural diuretic which is of some help in preventing swelling of the wrists, ankles and face due to water accumulation. B-6 is needed for the proper use of vitamin B-12, in digestion, red blood cell production, release of glycogen from the muscles and liver and for making antibodies for the immune system. It is needed for the making of nucleic acids and for balancing the relative amounts of sodium and potassium in the body. A quart of goat milk gives us 35% of the amount the average person needs.

Vitamin B-12 (cobalamine). This cobalt-containing vitamin is essential in the formation of red blood cells. We get 12% of the

RDA in a quart of goat milk, so we need to take enough meat, fish and especially liver to get the full amount we need. The best vegetable source is the edible alga, chlorella, available in health food stores. There is some in dairy products. B-12 is required for nervous system maintenance and the metabolism of foods. B-12 works best taken with vitamin C, potassium, sodium, iron and other B-vitamins.

Pantothenic acid. A quart of goat milk supplies nearly 1/3 of the RDA of this B-vitamin. This vitamin stimulates the adrenals, promotes cortisone production, takes part in cellular metabolism, stimulates liver production of cholesterol and steroid hormones, and helps keep the digestive tract healthy and working right.

Folic acid. This B-vitamin works with B-12 and C to help break down proteins. It is necessary, together with B-12 and iron, for the formation of red blood cells. Lack of iron, folic acid or B-12 will result in anemia. Folic acid is used in the production of nucleic acid. It regulates the appetite, supports the liver and is involved in cellular reproduction. There is so little folic acid in goat milk that one doctor recommends adding 100 micrograms of folic acid to every quart of goat milk used.

Biotin. We find 12% of the RDA for this B-vitamin in a quart of goat milk. Biotin helps form fatty acids, and works together with folic acid, pantothenic acid and B-12 in utilizing protein.

Choline. Over 1/4 of the RDA of the B-vitamin choline is supplied by one quart of goat milk. Choline can be manufactured in the body by the action of B-12 and folic acid on the amino acid methionine. Choline works with inositol to form lecithin and keeps fat from accumulating in the liver. Choline is needed for myelin sheaths of nerves, helps regulate the liver and gallbladder and helps protect the kidneys. Lecithin is a good choline supplement source.

Inositol. We get nearly 40% of the RDA of this B-vitamin in a quart of goat milk. It helps in the metabolism of fats and reduces blood cholesterol. Inositol is needed by brain cells. It protects the liver, kidneys and heart and is used for the growth and maintenance of cells in the bowel, membranes of the eyes and bone marrow. The body makes enough inositol to compensate for any deficiency in the daily diet.

Vitamin C. This vitamin is used in forming red blood cells, fighting infections, healing tissues and maintaining collagen tissue, which is used in skin, ligaments and bones. Vitamin C interacts with amino acids, folic acid, calcium, thiamine, riboflavin, pantothenic acid and vitamin A. The body normally holds about 5 grams of this important vitamin. A quart of goat milk has 23% of the RDA, which is 60 milligrams. Some people believe in taking from one to five grams of vitamin C. In that case, you would have to take supplements as found in natural health food stores.

Vitamin D. This is called "the sunshine vitamin" and only 6% of the average adult's need is met in by a quart of raw goat milk. It is important for you to understand that too much vitamin D can be toxic and too little prevents the intake of calcium and phosphorus we need each day. Vitamin D aids in synthesizing enzymes involved in calcium transport. It also normalizes the nervous system, heart action and blood-clotting mechanism. Vitamin D can be manufactured in the body by the action of ultraviolet light on a type of cholesterol as it passes through the blood capillaries just under the skin. The RDA is 400 I.U. and this should not be exceeded without consulting a doctor. Too much vitamin D can lead to calcification of soft tissues, walls of blood vessels and kidney tubules. Symptoms of overdose are frequent urination, nausea, vomiting, diarrhea, dizziness, weariness and weakness.

Vitamin E. There is none of this vitamin in goat milk, but because it is such an important vitamin, you need to get it in other foods. Natural sources include wheat germ, wheat germ oil, raw seeds and nuts, soybeans and cold-pressed oils. Vitamin E is a cancer-fighter, and when it is taken together with natural selenium, the anti-oxidant power of both are significantly increased. This vitamin prevents fats in the body from going rancid, and prevents vitamins A and D from going bad. Vitamin E stimulates cellular respiration of all muscles, including the heart. It increases stamina and endurance, dilates the blood vessels to the heart and protects red blood cells from attack by peroxide. It retards aging and appears to have a beneficial and protective effect on the sexual system. When goat milk is given to babies, the milk should be fortified with enough vitamin E to give them 4 I.U. per day. The adult RDA is 15 I.U., but many adults take 400 I.U. without any problems.

AN IMPORTANT REMINDER

I want to remind you that all vitamins work together with minerals and trace elements in building and maintaining your health and well-being. This means it is very important for you to make sure you are getting all the minerals and trace elements you need in your daily food regimen. If you are taking vitamin supplements when you are deficient in minerals, you are wasting both the vitamins and the money you spent on them.

One of the best natural mineral supplements I know is Whex, a dried goat whey powder which provides a mixed array of minerals and trace elements in easily-digested form. The following chart will show you what is in the Whex. If you don't get all the minerals and trace elements you need in your daily diet, this may be for you.

MINERALS AND TRACE ELEMENTS IN WHEX*

("Whex" is now called Capra Mineral Whey.)

(Can be taken by dissolving one or two heaping tablespoons in a glass of warm water or tomato juice.)

Element	Amt in 1 tbsp (20 gm)	Element	Amt in 1 tbsp (20 gm)
Boron	104 micrograms	Nickel	9.4 micrograms
Barium	62 micrograms	Phosphorus	122 milligrams
Calcium	58 milligrams	Sulfur	17 micrograms
Cobalt	2.8 micrograms	Selenium	28 micrograms
Chromium	10.4 micrograms	Silicon	9.4 micrograms
Copper	19.8 micrograms	Silver	10 micrograms
Iron	132 micrograms	Tin	76 micrograms
Potassium	50 milligrams	Strontium	32 micrograms
Lithium	19 micrograms	Titanium	9.4 micrograms
Magnesium	24 milligrams	Vanadium	2.8 micrograms
Manganese	1.8 micrograms	Yttrium	1.0 micrograms
Molybdenum	9.4 micrograms	Zinc	54 micrograms
Sodium	84 milligrams		

I gave the amount that is in one heaping tablespoon, but you can mix yourself a cup or glass of Whex with two heaping tablespoons and take it as many times as you want to during the day. This is a food, not a drug.

This product (Whex) is in Dr. Jensen's Food Blend, a whole food supplement that has helped many patients and others who use it boost their level of wellness.

It is important to know that vitamin C can form chelates in the bowel with many known metals (such as zinc, copper, iron, magnesium, etc.), which means the chelated metals are prevented from being assimilated through the bowel wall and are eliminated. **Don't take vitamin C at the same time you take your mineral supplement.** It may work well for you to take your minerals and trace elements during the day and take a large dose of vitamin C just before bed.

There is a great advantage to taking these minerals together in food form. The body can then select out the ones it needs the most and assimilate and use them. Goat milk and goat milk whey are wonderful sources of minerals and trace elements. I would encourage you to try them. They have done a world of good for my patients.

Goats are playful, kind, loving and climbers. These elements are necessary as a source for good milk we need in our body. Consider the source for your food and especially milk. Did you ever see a cow on top of a shed?

CHAPTER 7

GOAT MILK FOR INFANTS AND CHILDREN

Before an infant is born, it lives on the nutrients drawn from the mother's blood. If the mother's diet carries all necessary nutrients, the fetus will be well nourished. If the mother's diet is deficient, the baby will be born with the same deficiencies as the mother. The children of chemically-addicted mothers are often deficient in a number of minerals and vitamins. The vital beginning for a child's organs, glands and bones lies in the health of the mother during and after pregnancy, as she enters into nursing the new baby.

Healthy mothers can begin to breast feed their infants and expect good health and growth. It is the better part of wisdom for an unhealthy mother to consider another way.

Besides the mineral deficiencies in the newborn due to chemical abuse by the mother, we find that many a mother hasn't taken care of herself sufficiently to avoid various nutritional deficiencies in the baby. It is better for such mothers to consider using goat milk to help the infant catch up as much as possible with its nutrient deficiencies.

Goat milk is one of the best possible foods to give an infant who has been deprived of nutrients for these and other conditions developed during the mother's pregnancy.

Other babies suffer at times from indigestion, diarrhea, vomiting and colic. Some have been overfed, others may have suffered from sugar fermenting in their systems. Sometimes the mother is nervous, cranky, fussy or too busy to take care of her baby. Goat milk soothes the digestion and calms the digestive tract. A mother who is disturbed, miserable or unhappy will find that her milk is adversely affected and is not suitable for her baby. She should consider giving goat milk to her baby at such times.

GETTING OFF TO THE RIGHT START

Colostrum, the "first milk" secreted for about three days after an infant's birth, stimulates the first peristaltic motion in the baby's bowel and establishes the friendly bacteria in the bowel. Colostrum also supplies the newborn with antibodies and immunoglobulins to protect it from sickness and disease until its own immune system is activated.

If acidophilus bacteria—the friendly" bowel bacteria—are lacking in the infant's bowel, which can happen if the child is not fed properly in the beginning, we can add it to the diet. Health food stores carry an acidophilus culture that can be bought and added to the baby's milk until the intestinal flora are normalized. Whey will feed the acidophilus best, after the acidophilus has been given daily for a couple of months. Without good flora in the bowel, the infant may have gas, a dry bowel, diarrhea and other problems.

The whey used to feed the bowel should be goat whey and may be liquid or dried. The dried whey is more concentrated.

If the baby can't take the acidophilus culture by mouth, it can be administered by a baby syringe into the rectum.

A clean, regular bowel is important to a child's health.

BOOSTING THE EFFECTIVENESS OF GOAT MILK

Goat milk for the newborn is an effective substitute when "normal" mother's milk is not available. The goat is a healthy, serene animal, not subject to most of the diseases, conditions and emotional states that can make breast feeding impossible or impractical. Its fat molecules are 1/5 the size of cow milk molecules, which makes goat milk easily digested by nearly all infants.

Goat milk is lacking in two vitamins and one mineral important to the development of infants and children. These are folic acid (a member of the B-complex group), vitamin E and iron. Folic acid is needed to break down proteins, form hemoglobin, make nucleic acid (DNA), stimulate liver activity, regulate appetite and assist in the production of hydrochloric acid for digestion. Vitamin E is an antioxidant, protecting against rancid fats and oils, and guarding against tissue damage by free radicals. In the latter role, it is anticarcinogenic. Vitamin E stimulates cellular respiration, protects capillary walls, prevents blood clots from forming, protects red blood cells and aids in feeding the cells. It is believed to slow down the aging process. Iron is needed by red blood cells to carry oxygen to all tissues. It is also needed to oxygenate muscle tissue and by enzymes involved in protein metabolism.

Vitamin E in the amount of 4 I.U. (International Units) must be given daily to babies during their first year, since it is lacking in goat milk.

Because goat milk is deficient in folic acid, deficiency can lead to anemia, diarrhea and other problems, it is best to add 100 micrograms of folic acid to each quart of goat milk used, whether by infants or adults.

All milk is low in iron, including goat milk. At the point where the infant may need a little iron supplement, blackstrap molasses can be added to the milk.

Dr. Charles Parry, nutritionist, says that goat milk with 5-to-10 milligrams of niacinamide will often bring infant diarrhea under control within 24 hours.

REQUIREMENTS FOR BABIES AND CHILDREN

	0 - 6 mos.	1 - 3 yrs.	4 - 6 yrs.	7 - 10 yrs.	11 - 14 yrs.	15 - 18 yrs.
Iron	10 mg	15 mg	10 mg	10 mg	18 mg	18 mg
Folic Acid	0.5 mg	0.1 mg	0.2 mg	0.3 mg	0.4 mg	0.4 mg
Vitamin E	4 IU	7 IU	9 IU	10 IU	12 IU	15 IU

I have seen goat milk do wonders for babies and growing children. Its nutritional superiority soon takes care of most of the deficiency problems developed in the child prior to or after being born. Goat milk has a much greater range of minerals and trace elements than most foods, and only needs boosting in nutrients like folic acid, vitamin E and iron.

CHILDREN AND CALCIUM DEFICIENCY

Dr. Philpots of Melbourne, Australia, after 15 years working in the Children's Hospital says: "I have seen children at the age of 18 months to 2 years, with their teeth decayed down to the gums and badly abscessed, caused by incorrect diet. A complete lack of vitamin foods, extended over a period of years, produces sickly children with bad teeth."

Frailty indicates ill health and disease in children. Look for the cause in the diet. If it is there, remove it. Leg aches and other "growing pains" are often a sign of a calcium deficiency.

Candies, puddings, cakes, pies, fried foods, other pastries, gravies, pickles, jellies, preserves, sweetened fruit and drinks, devitalized and sweetened cereals and white flour products are among the taboo foods for raising healthy children. Such foods promote frailty and sickness.

Foods which should be fed to growing children are:

- ☺ Clean (raw) unpasteurized goat milk.
- ☺ All vegetables except dry beans and dry peas, which are difficult to digest.
- ☺ Fresh fruits other than bananas (that have to be ripe and mellow).
- ☺ Cooked fruits sweetened with honey, dried fruit or pure maple syrup.
- ☺ Cottage cheese, buttermilk and clabbered milk.
- ☺ Chicken, lamb and occasionally other meats except pork. (This does not apply to the vegetarian regimen.) (Children under 7 years are best without meat.)
- ☺ Whole-grain bread with raw butter.
- ☺ Whole-grain muffins or biscuits.
- ☺ Whole-grain cereals to be eaten dry with honey and maple syrup. Revived dried fruit may be added.
- ☺ Whole-grain cakes and cookies made with molasses, maple syrup and raw sugar. Serve only occasionally.

SOME BENEFITS FOR BABIES

Raw goat milk usually contains natural fluoride in a sufficient amount to reduce or prevent tooth decay as the baby's first teeth begin to form. Fluoride hardens bones and teeth, and in the case of teeth, prevents cavities.

Calcium, phosphorus and magnesium are generously supplied in goat milk, to build strong bones and teeth.

There is plenty of food sodium and potassium to take care of metabolic acids in the stomach, bowel, joints and muscles. Goat milk has the right trace elements to meet the needs of specialized proteins and enzymes being formed in the infant's body.

Goat milk works its wonders through its cleansing and rebuilding activities at the cell level.

WEANING THE BABY

For mothers who have reached the point where it is time to think about weaning the baby, I suggest this formula to use along with mother's milk. Goat milk is the ideal food for weaning since it is nearest to mother's milk in composition and properties. It is easy to digest and it puts on weight. Goat milk is naturally homogenized. You should consider adding 100 micrograms of folic acid to a quart of goat milk before taking a little of it to make the formula.

WEANING FORMULA

1-1/2 ounces raw goat milk
3 ounces distilled water
1/2 ounce pure cream
1 tsp sugar of milk

This is the formula I have recommended for weaning the baby. When the time comes, it is well to add a little barley gruel, and gradually raw vegetable juices (strained and diluted with water or goat milk, at first).

You may also use a little broth made of carrots, onion and parsley finely cut and cooked in a little water. Strain and use only the broth.

Fruit juice made from reconstituted, unsulfured dried prunes, figs or apricots can be diluted further with water and given to the infant. By reconstituting dried fruit, I mean boiling it for a few minutes in water, then allowing it to cool.

MENU SCHEDULE FOR WEANING

6 am: Milk formula (above)
8 am: Fruit or fruit juice
10 am: Cereal and milk formula (above)
2 pm: Milk and steamed or pureed vegetables
4 pm: Broth or celery and parsley soup
6 pm: Cereal and milk formula (above)
8 pm: Fruit or fruit juice
10 pm: Milk or milk formula (above)

Distilled water may be used for the infant first and gradually reduced. Vary as weaning advances. After a couple of weeks, depending on the child and the reactions, use straight goat milk or raw milk (or milk substitutes). If goat milk can't be tolerated, try cow's milk (raw). Begin introducing other foods at this time. Include foods high in silicon, calcium, fluorine and iron, among others. Emphasize these minerals most. As we have said, it is best not to give the youngster meat until about age 7.

SUPPLEMENTAL DIET AT WEANING TIME

1. Orange juice (tree ripened only), with some pulp, diluted with distilled water.
2. Fresh tomato juice, grape, papaya, pineapple and apple juice.
3. Figs and prunes, raisins and dates—soaked for about 12 hours (revived) and then pureed or mashed. Also their juices.
4. Any fresh, uncooked fruit, pureed, mashed or cut into tiny pieces to introduce the child to the new taste and texture.
5. Any whole grain cereal, including brown rice, millet and yellow cornmeal that has been steamed in a double

boiler over very low heat several hours or thermos cooked overnight. Cream whole grain cereals. Other good cereals are oatmeal, cream of rye and creamed whole barley.

6. Any fresh vegetables in season—steamed, mashed or cut into tiny pieces or pureed in a blender. It is best not to have more than two varieties at a sitting. Use very small quantities at first. Add broths made of carrots, parsley, etc.
7. Replace part of mother's milk with goat milk gradually to assure a natural and healthful break. You can add carrot juice and other fresh vegetable juices to goat milk any time. My favorite is goat milk and carrot juice, half and half.
8. Do not flavor baby's food. The natural flavor of foods is adequate and healthier. If the child displays a dislike for a cereal, flavor with a little fruit juice concentrate, unheated honey or dried fruit. **Do not use salt or condiments.**

FURTHER SUGGESTIONS FOR WEANING TIME

At or near weaning time, the amount of goat milk may be reduced by substituting soy milk powder mixed with water for a consistency similar to milk. This may be used in weaning the baby.

When the infant is well on the way to being weaned, add rice water, barley gruel, soaked milk and berry juices (without seeds). Mash young tender carrots with goat milk, especially when the youngster is older and starts teething. Add fresh figs and soups made of carrots, parsley and celery. Add some juice of sweet fruits—fig and prune. Introduce raw vegetable juices, soft-boiled eggs, tender omelets and souffles, spinach toast (stale bread), fish-bone broth for bone development and teeth building, if not vegetarian.

GRADUATING TO SOLID FOODS

You can make your own baby cereals if you choose. It is time consuming, but not difficult. Simply cook the whole grain cereal until done, put a little in your blender, add just enough warm water or goat milk to make a thin, smooth, semi-liquid cereal. Mashed banana will be one of the early semi-solid foods a baby likes.

Don't use sugar or salt on your children's foods. They don't need either. To sweeten, use fruits like raisins, figs or dates (cut up fine) to mix with the cooked cereal. Or, blenderize just enough fruit with the cereal to give it a pleasant, sweet taste.

Start your children on natural foods, and they will want only natural foods as they grow up.

Don't be afraid to fast a child or let him miss a meal or two. Children seem to know when they shouldn't eat, although we have to be careful here not to let them have candy or junk food between meals. Between-meal snacks must be carefully monitored and should be limited to juices and fruit. If your child is not growing properly, see your doctor.

GOAT MILK FOR GROWING CHILDREN

From early childhood up through the teens, goat milk should be an important part of your children's food regimen. Remember to supplement the milk with folic acid (needed to build a vital bloodstream) and iron (use blackstrap molasses). A baby in the beginning of life has enough iron to take care of the body's needs for about a year after birth. Beyond that time frame, the child should be taking foods that have iron—greens and perhaps finely-pureed meat.

As I have mentioned, goat milk compares favorably to mother's milk. In contrast, cow's milk has elements suited to the building of heavier bones, hooves and a hairy hide. The digestion of the calf and its larger liver allow it to handle the larger fat globules in the cow milk, but a human infant doesn't have that same capability. Cow milk doesn't compare with goat milk in ease of digestibility and assimilation. The small fat globules in the goat milk are easily handled by a human infant's small liver.

The goat is a very active, limber, nimble animal, a sign that it is a sodium animal. Goat milk often (but not always) is richer in sodium than cow milk. It depends upon feed. But, the sodium in the goat milk is higher evolved than that in cow milk. Sodium is needed in the infant's body to keep the stomach sweet and normal in its hydrochloric acid secretion. The hydrochloric acid is needed to digest protein and prepare calcium for assimilation. Sodium is needed in the bowel wall to neutralize bowel acids and in the joints to make them pliable and limber. Children are usually very limber.

Milk is better utilized if it is taken together with other foods at mealtime or snack time. A child should not drink milk rapidly and in large amounts at one time. A straw may help slow him down.

Goat milk is naturally homogenized so that the cream doesn't separate from the milk and rise to the top as cow milk does. There is less cream in goat milk than cow milk, but, after all, we are not trying to build too fat a baby, and we don't want to overload the liver with fat. It is better to have a healthy baby, and a healthy baby is not overweight.

I want to emphasize that goat milk is a superior food for all children, not just children with deficiencies or health problems. We should realize that any food that benefits people with health problems will help prevent health problems in normal people. Goat milk is a wonderful food for people of all ages.

FOOD COMBINING AND CLABBERED MILK

Milk is more easily digested and taken care of if it is mixed with other foods. In a heavy-acid stomach, milk should definitely be mixed with other foods and, many times, diluted, so it doesn't curdle in the stomach too fast. If the milk does not agree, have the child sip it slowly or take it through a straw, so the stomach can handle it more easily. The child should not drink milk rapidly and in large amounts at one time.

For the child, the best foods to combine with goat milk are vegetables or fruits, and the fruit can be liquefied, if desired, with the milk. All stewed dried fruits are good with milk, and should be used if fresh fruits are not obtainable.

Clabbered goat milk is an easily-digested food that can be taken by old and young alike. The changing of milk into the clabbered form involves an enzyme reaction like that which happens in the stomach with hydrochloric acid. Clabbering the milk before it is eaten gives the stomach an easy way to digest milk proteins.

To make clabbered milk, all utensils must be clean. Heat 2 quarts of raw goat milk to lukewarm. You can add some powdered goat milk **before** the milk is heated, up to a half cup. After the milk is heated, add 1 cup of it to 1 cup of cultured buttermilk or yogurt, stir well, then add back to heated goat milk. Place in the oven with only the pilot light on, or in some other suitably warm place where it can set undisturbed. The time it takes to set may range from 4 to 8 hours, depending on the culture and the temperature (90 to 115 degrees F). When it is ready, cover the pan and refrigerate.

All digestion of the milk does not take place in the stomach, but the coagulation of the milk begins there, and if milk is taken too fast and without the mixture of other foods, coagulation is very difficult and, after it leaves the stomach, the milk cannot be completely digested by the other gastric juices along the intestinal tract.

Many think milk is very constipating. Cow's milk is exceedingly constipating but goat's milk or cheese is not constipating because goat's milk contains lactic acid in abundance. It also contains chlorine, sodium, potassium and calcium salts. Goat's milk is exactly what children need. Do not boil goat's milk because you will kill the vitamins and enzymes. Feed children the natural product; never interfere with the Almighty's plan. Goat's milk is rich in chlorine and fluorine—the very elements that reduce bacteria in the body.

Most children who take goat's milk have better bowel movements, and they seem to thrive better than those given cow's milk. We have found, in many cases, that eczema and allergic conditions, asthma and many catarrhal ailments of children have responded by merely using goat's milk instead of cow's milk.

CHAPTER 8

THE BEST FOOD FOR MAN

Without doubt, goat milk and its products are among the most important foods that can be produced or purchased. Goat milk, cheese, buttermilk and whey are important foods because they come from the cleanest of all animals. The eliminative organs are almost perfect, and their disposition favors the health qualities of the milk.

The atoms of goat milk have a higher rate of vibration and a superior degree of sublimation. The goat stands closer in evolution to man than any other non-primate. The disposition of the goat is friendly. Its milk chemistry is superior and stands closest to mother's milk than any other milk.

Goat milk is even superior to mother's milk, because it is higher in fluorine, sulfur and other elements of great value to human health. It is reconstructive, sanative, even curative to certain patients, children, babies, especially all who are low in those food elements in which goat milk is high. Goat milk, unlike cow milk, does not cause phlegm in children or adults who use it. It is a fact that states of mind have their effects on the milk, on the secretions, on the chemistry of the body and on the chemistry of the milk. A scientist analyzing the body chemistry found that he could obtain 40 different chemical products due to the influence of various emotions. Goat milk is produced under favorable states of emotions and dispositions because of the friendly nature of the goat.

The goat is as gentle as a loving daughter. It is because of the high evolution of the goat that her milk, in the past, has enjoyed such a high reputation for its curative, recuperative, blood-building power, to ailing babies, children, even to ailing men and women who were low in vital vigor. It is true that goat milk does not cure all diseases, that it does not raise people from the grave. But it is also true that it often cures where other therapies fail to cure. This has been proven to us thousands of times, and to thousands of others long before we came on the arena. Deficiency diseases or diseases due to imbalanced diets are often cured by goat milk, especially in people low in the minerals needed to sustain health.

The author of this article has had ample opportunity to test out and recommend goat milk or goat milk products in some form to thousands of ailing children and patients. So remarkable has been the result that it has often been almost unbelievable what goat milk will and can do to certain types of patients, providing such patients are not overfed nor underfed, nor given so many drugs that their power of resistance to disease has become too far depleted.

GOAT MILK IS CURATIVE

Often we hear it said that "milk is milk." While there may not be a great deal of chemical difference between milk and milk, yet, one milk may cure while the other milk may fail to cure, and no doctor, no chemist, knows why.

THE NUTRIENTS ARE SUPERIOR

That delicious milk that comes from a young, high-bred Saanen, Toggenburg, Alpine or Nubian goat is alkaline, healing, sweet and remarkably high in nutrients. Milk has its own chemistry, taste, consistency, specific gravity, bacteria, alkalinity, acidity and other qualities. Goat milk differs from cow's milk as every food chemist knows. Goat milk has less protein, fat, ash and less calcium; but it has more fluorine, silicon and other trace elements.

Goat milk has a superior cholesterol and a finer brain phosphorus than cow's milk. The constituents in milk vary to a certain extent in different samples of milk, but never to any great extent. The water content of milk never varies much from 85 to 87% and the solids seldom run less than 13%. Goat milk is nearly germ proof, but cow's milk often has bacteria in abundance. The chemist can see microscopic wigglers in raw cow milk by the tens of thousands, but in raw goat milk there is never such an army of wigglers.

The taste of goat milk is superior because goats are choosier about what they eat than cows. If you give a cow an onion, the cow will eat it, and the milk will pick up the flavor of onion. But the feed of the goat is superior.

The cholesterol found in goat milk is of high value for the marrow of the bones, for the communication fibers in the brain, for the nerve substance, for the lymph in the ears, for the spinal ganglia, for the nerve nets, for the brain cells whose number exceeds one billion, according to the investigation of that great cell student, Professor Robinson.

GOAT MILK IS AN EXCELLENT BRAIN FOOD

No better food can be found for man's wonderful thought machine (the brain) than goat cream or goat milk products in some form. But, after all, the chemist in his laboratory cannot determine all the factors in the case. His analysis deals only with chemistry or with the visible factors. The invisible factors evade his analysis and his instruments. Why the cholesterol or goat cream has an affinity for brain and nerve matter, no chemist can explain. Nor can the food chemist determine by his analysis why cow cream and cow butter has an affinity for the neck, hips and lower bowels. Nor can the doctor, nor the chemist understand why goat milk and cream should agree with certain types of babies, but not with other types of babies.

The goat has been called "the poor man's cow." But she is as much the rich man's cow, for there are wealthy people who pay fabulous prices for goats and keep them for their milk. The health of an animal has its effect on the quality of its milk. The goat is a remarkably healthy animal.

THE MILK IS FROM A HEALTHY ANIMAL

The goat's eliminative organs are usually perfect, so that anything that is detrimental to her well-being is eliminated at once. Her "instinct" of self-preservation is exceedingly keen.

People say that goats eat tin cans. Many people fail to feed their goats, so the poor animals nibble at the labels on the cans, and the goat owner, or neighbors, think that goats eat tin cans. Goats have a fancy taste and are clean in habits. They have a reliable nose for grass, bark, leaves and other feed. They know by instinct what to eat and what to leave alone, unless they are starved or forced to eat wrongly. They have a normal appetite.

Fresh goat milk is one of the cleanest of foods, because the goat is cleanly in eating habits. Indeed, the chemist knows of no cleaner food than fresh goat milk, coming from a healthy young goat. But this is not true in regard to cow milk, nor is it true in regard to meat, fish and poultry.

The bacteriologist or the microscopist knows that there are germs, bacteria, impurities and toxins in almost everything we eat, and he knows that goat milk, if fresh, is the safest food for man or for a baby. Indeed, the Great World Builder, God, knew His business when He created the goat. Goat milk and goat cream, if fresh, are always alkaline and germ-free.

We have learned that a well selected goat milk diet is imperative in many ailments and diseases such as: nervousness, brain fag, neurasthenia, scrofulous ailments, first stage of tuberculosis, blood diseases, pneumonia, intestinal ailments, pleurisy, arthritis, osteoporosis and many others.

Goat milk feeds the skin, the hair, the nails, the teeth and the bones, as well as the brain, the nerves and other structures of the body.

GOAT MILK FOR HEALTH AND BEAUTY

Not that goat milk and cream is an absolute cure for all diseases. It simply aids up-building, beauty building, blood building by supplying the food elements that are essential, also the cholesterol and the food salts that are needed for curative purposes.

Because fresh goat milk contains all of the 16 chemical elements needed by man in almost perfect proportion, it is one of the best foods known to man. Health-minded people should study the chemistry of goat milk, its products, use and combination with other foods for health-building purposes.

BEST AND CLEANEST FOODS FOR MAN

Fresh goat milk from young goats,goat cream, goat cottage cheese, fruit gelatins, baked bananas with goat buttermilk, steamed carrots with goat buttermilk, baked almond nuts, fresh eggs from healthy young hens, delicious eggs prepared with goat cream and cooked in a double boiler, almond meal muffins, baked fish roe, goat ice cream, well-cooked oatmeal with goat milk cream and raw well-beaten egg whites with a little goat cream.

Such foods can be recommended to anyone whether healthy or not, or even in any season of the year, in spite of the fact that people are different in chemical type and in spite of the fact that "one man's meat is another man's poison," as doctors know.

For brain and nerve nutrition, we should add green turtle broth, fish roe or fish eggs and goat cream together with goat milk and goat buttermilk.

Probably the best bread substitute is Ry-Krisp.

For more rapid growth in children, add high potassium foods such as goat milk, greens and bananas.

For flesh building in lean people, use goat milk and goat cream, sour goat milk, grated carrots, baked bananas and the strippings of the milk at milking time. Breathing exercises after eating and sleeping much are also highly desirable for flesh building in lean persons.

It should be remembered that goat milk and goat milk products vary in chemical contents, even in the same goat, at different seasons of the year. For instance, the strippings contain from 8 to 10 times as much brain and nerve fat as the foremilk.

Goat milk contains the mineral salts as compounds of phosphates, chlorides, citrates also fluorides and bicarbonates. The mineral ash of any milk does not always represent all of the minerals because of the processes of incineration and oxidation. However, two hundred samples of goat milk analyzed proved conclusively that the formula given in this chapter is technically correct.

The goat deserves a place in our lives. Its milk and milk products are needed in some form in every home, hospital and convalescent facility.

CHEMICAL PROPERTIES OF GOAT MILK

As nutrition specialists, we have formulated diets for thousands of ailing people, studied food chemistry and recommended goat milk in some form for adult sufferers for many years with excellent results. Thus, we know that goat milk products bring results to ailing people and greater vigor to all. The chemical properties of the goat milk have a great deal to do with this success.

We shall here consider chemical properties of goat milk, goat cream, goat butter, goat cottage cheese and perhaps also goat buttermilk and goat cheeses, if time permits.

Goat milk, as drawn from a healthy goat, is amphoteric, which means that it reacts acid with blue litmus paper and alkaline with red litmus paper. Let the milk stand about 4 hours and the milk sugar slowly is converted into lactic acid. This chemical process increases on a graduated scale until lastly the acidity reaches a maximum in 72 hours at a medium temperature.

This acid-forming process can be prevented by adding preservatives to the milk, such as sodium carbonate, boric acid or other alkaline preservatives. But if the milk is drawn from the goat in kid fashion, no preservatives are needed.

NOT ALL MILK IS WORTH DRINKING

Man, however, has turned his back on heaven and follows his own commercial notions. The results are that the commercially-processed milk, cream, cheese, butter, buttermilk and other milk products that man uses in his home for babies, children, himself and other family members are acid-forming and therefore, may create trouble in the chemical laboratory of his and their stomachs. Lastly, his children become sick, and he himself is the cause of his children's and his own suffering. He calls it destiny and often, in his ignorance, charges it up to God. But, in each case, he pays heavy doctor bills and hospital bills.

The milk may be pasteurized, to kill the milk wigglers, and man calls it efficiency. But he does not know that he has altered the chemistry of the milk, nor that his children will suffer from bone disease, skin ailments and poor teeth later on. The health of his family fails; love takes wings to itself and flies to Siberia and he ends up in the divorce court. He becomes discouraged and blames politics or perhaps the current president for his calamity. But he forgets that milk should be used as it comes warm and foaming from the young and thrifty goat.

GOD'S PRODUCTS ARE THE BEST

Science and art cannot improve on the Almighty. Science often walks on crutches and so do its victims. We must eat correctly or our doctors fail to cure.

The yellowish creamy globules in goat cream constitute that which the chemist calls cholesterol. This fatty substance is the best food for nerves, lungs, marrow of the bones, brain cells, optic centers, eyes, ear mechanism and for general lubrication of the human machine. This fatty substance in goat cream is superior to tonics for tired nerves, for crying babies, hysterics, brain workers, tired school teachers, weary lawyers or for anyone whose nerve batteries are burned out.

FRESH GOAT MILK IS ALKALINE

As the milk comes fresh from the goat, germs are seldom present. Bacteria thrive in a medium of acidity, but not in a medium of alkalinity. Fresh goat milk, cream, butter and buttermilk are always alkaline and curative, but acid milk products are not, as every food chemist knows. Kids, kittens, colts and babies obtain their nutrition direct from God's nutrition factory, according to the wise plan of the celestial chemist, God.

Milk and milk fat differ in different goat breeds and even in the same goat in different seasons, on different feed or pastures. Stall-fed goats have leaner and bluer milk. Spring pastures favor fat building. Calcium phosphate is found mainly in cottage cheese. Goat cottage cheese is a bone builder and teeth builder in growing boys and girls and for certain almost boneless types of people, having flabby, lymphatic, watery tissue. Give such people goat cottage cheese. The casein of goat milk contains about 80% of the entire proteins found in the milk, being present to the extent of 3%. The cheese of milk is soluble in an alkaline medium. If there is a deficiency of sodium in blood, tissue and secretions, a milk diet leads to arthritis, stiffness, arteriosclerosis, poor hearing, weak eyesight, calculus, urinary ailments, catarrh and "old age ailments." That is why a milk and cheese diet calls for sodium food. A high-sodium diet cures calcic ailments and diseases, but not other rheumatic ailments. There are many different kinds of arthritis, rheumatism, hardening.

Milk albumin is contained in goat milk to the extent of about 6%. Milk albumin resembles egg albumin and is soluble in water, and more or less, metabolized by sodium chloride, also by other chlorides. Food chlorides always favor the digestion and assimilation of albumin contained in protein foods.

SOME INGREDIENTS OF GOAT MILK

Other chemical substances present in goat milk, such as: globulins, fibrin, caseose, peptones, gelatin, galactin, amino acids also certain decomposition products such as: ammonia, creatine, urea. The fat in goat milk has 10 fatty principles, namely: glycerides, caprinin, caproin, caprylin, butyrin, butin, myristin, stearin, palmitin and olein.

The most important fats in goat milk are olein and myristin, as those fats favor nerves and brain cells, mainly. This fat is a highly-evolved animal fat, a fat that is unknown to the entire vegetable creation. It is believed that it contains "mind stuff," as one great chemist called it.

The milk serum in goat milk contains nearly 13% of solids: casein 3%, milk sugar 4.5%, fibrin, a trace; galactin 0.20%, with lacto-globulin; albumin 0.60%. Casein, galactin, albumin, lacto-globulin and fibrin contain nitrogen.

The ash in goat milk amounts to about 0.7%. In this ash, the chemist finds: potassium oxide, chlorine, phosphoric pentoxide, sulfur trioxide, iron oxide, magnesium oxide, calcium oxide, sodium oxide, also a sprinkling of fluorine, manganese, silicon, iodine and, at times, arsenic, aluminum and copper.

The arsenic and aluminum are considered foreign by prominent food chemists.

The water content in goat milk seldom exceeds 87.3% though it may vary slightly. Indeed, one chemist said that a man is mainly a bucket of water with a handful of minerals added. That is the material side of any man, but not the mental nor the spiritual. Outside of mind and soul man is nothing but mud and water.

DIFFERENCES IN TYPES OF MILK

Chemically, there is a considerable difference between milk and milk. Chemists know goat milk from cow milk or from human milk by the specific gravity alone. The specific gravity of goat milk averages 1.0305 while cow milk averages 1.0315 and human milk averages 1.035. There is a slight difference between the contents of water, protein, fat, ash, albumin, casein, galactose, also between the ash of the various milks. Thus, a chemist can always tell whether it is goat milk, human milk, dog milk, cow milk or any other milk by the different chemical properties prevailing in the various milk. Goat milk is nearly germ-free, even when raw.

The analytical process of milk is such that the ash contents are not scientifically accurate. Molecules may be altered, parts lost, chemistry incorrectly estimated or there may be waste or parts may be lost through oxidation or incineration. Yet chemists always agree that quantitative chemical analysis is an ideal chemistry. Many prominent chemists know from hundreds of quantitative analyses that the mineral contents in goat milk are represented, on an average, in proportion as follows: Calcium, combined with proteins, 5.13; calcium citrate, 23.01; tricalcium phosphate, 8.70; dicalcium phosphate, 7.39%; magnesium citrate, 4.01%; dimagnesium phosphate 3.70%;

potassium citrate, 5.15%; dipotassium phosphate, 9.18%; mono-potassium phosphate, 12.71%; potassium chloride (KCl), 9.14%; sodium chloride (NaCl), 10.60%, calcium fluoride, silicon oxide and manganese oxide, all at 1.86; iodine appears in traces; also, one or more of some foreign elements, in traces.

The bitter taste of goat milk, when there is any, is due to hircic acid. If the milk comes from high-bred goats, there is no bitter taste and no "goat odor."

The foremilk is higher in water and lower in butter fat, while the strippings contain less water than 80.17% and more fat 10.36% or nearly 8 times more cream than the foremilk contains. But albumin, sugar and ash do not vary in the foremilk nor in the strippings. A fat person should never eat nor drink the strippings, but a lean, dry patient needs the strippings.

The colostrum is the raw milk after the birth of young, within the first three days. Colostrum is oily, pungent, laxative, cleansing to the alimentary tract, yellow in color and high in immunoglobulins (16.36%), also high in nitrogen, potassium, chlorine. People who suffer from stubborn constipation should take colostrum pudding, at times, to favor intestinal lubrication. This laxative comes from God's laboratory to favor the alimentary tract of young kids.

MILK AND FERMENTATION

The lactic acid bacteria are the chief causes of milk fermentation, but there are also other species of bacteria that flourish in various milks, especially if that milk has been exposed to warmth, light, air. Quickly soured fresh goat milk or buttermilk, made from fresh goat cream, contain a great amount of lactobacillus acidophilus. Such milk, together with baked bananas, is beneficial for one kind of stomach trouble, constipation or rheumatism, but not for other kinds of ailments. There are many causes of stomach trouble. Many babies suffer from lactic acid indigestion, diarrhea, vomiting, colic. The baby may be overfed, or suffering from sugar-fermentation.

Goat milk is sometimes superior to mother's milk, because goat milk contains food elements that mother's milk does not contain. The goat also has a different disposition than an angry, fussy, nervous person.

The goat is a serene creature, not subject to hysteria, fear, disappointments, neurosis, nervousness, temper, overeating, passion or jealousy. A woman may be a drug- or alcohol-dependent person, but a goat is not. This is why goat milk often cures a crying, sickly baby. If a baby is not properly fed, how can it be well? If a woman is nervous, sick, miserable, fussy, how can she be a loving wife? If a husband is a dyspeptic, how can he be an honor to his country or a delight to the angels?

Give children alkaline goat milk, alkaline goat cream, alkaline goat cottage cheese, alkaline goat buttermilk and keep their bowels open, and Oklahoma storms in the homes will never raise the roof.

Christianity does not thrive on an acid stomach, and the love of a dyspeptic or angry wife is a sour compound. To fall in love with her or to fall in love with a pessimistic dyspeptic is a grave mistake.

THE CREAM IS MORE DIGESTIBLE

Natural separated cream contains, according to Koenig, 68.82 parts of water, 3.76% proteins, 22.66% fat, 4.23% sugar, 0.53% ash, 31.18% total solids, 8.42% solids not fat, as proven by 40 analyses of separated cream. The fat particles in goat cream are five times smaller than those in cow milk and much less hard on the liver.

It seems that centrifugal separator cream, both light and heavy, has a different chemistry, as proven by 18 different analyses made by Leach, a prominent chemist. Heavy cream contains 51.68% water, 42.02% fat, no sugar, no ash but 48.32% total solids and 6.30% solids, not fat. Light cream contains about 25% more water than heavy cream, and about 28% less butterfat. Heavy cream contains almost 3 times as much butterfat as light cream. Centrifugal separator cream contains more than twice as much cream fat as natural separation cream does, because all the cream is not taken out of the milk.

U.S. Standard cream should and must contain 18% butter fat.

Crude gelatin is often used in cream to make people think that they buy heavy cream.

Anyone who keeps a goat or two for family use can obtain milk, buttermilk, cream, butter, colostrum, goat cottage cheese, mesost, homemade cheeses and other health-building foods.

Indeed, every housewife who is interested in her health, comfort and beauty, or in the long life of her husband or in the health of her children, should keep a few family goats or else let some kind neighbor keep them for her and supply her with health-building, beauty-building food, to increase her feminine grace.

A family should live closer to God, nature and the goat, than to the can opener. When God created Adam and Eve, He created no can opener for Eve, because he knew that if she did not feed Adam rightly, their marriage would end up in the divorce court. It does not pay to eat adulterated foods. Indeed, a few young thrifty hens, a few berry bushes, several gentle goats, a dozen fruit trees and nut bushes and a plot for vegetables would make almost any household independent. If you and we were wise, we would raise our own food and let food manufacturers go to Jupiter and read ***Robinson Crusoe***. Greedy men are not needed in the scheme of evolution.

THE BEST CHEESE IS MADE FROM THE BEST MILK

The chemical processes of ripening cheese made of goat milk are not well understood, not even by expert food chemists.

They may understand the maturation of the nitrogenous bodies, of the albuminosis, of the peptones, the amides, ammonia, fat, water, salts, lactate, paranuclein, moisture and its action upon enzymes and on certain known and unknown bacteria. They may know what favors proteolytic changes and the digestibility of certain varieties of cheese, but not all cheese varieties. Nearly all varieties of goat cheese contain approximately the same relative compounds, elements, salts, etc., but the enzymes, the microorganisms, the stages of maturation are not understood by cheese makers. There are so many different kinds of microorganisms at work and different germ varieties have a different chemical effect upon the chemistry of the various goat milk cheeses or, in fact, on any cheese. Cheese adulterants are also used, but an honest chemist is not interested in adulteration.

U.S. Standard cheese should contain not less than 50% butter fat.

If fresh foaming goat milk is separated and the cream placed in a cooler of 60-65 degrees F. for 24 hours and then churned and the buttermilk placed in a cold refrigerator for daily use, it is an excellent food for brain, nerves, ganglia, nerve plexuses, also favorable for general lubrication of the body. Placing the cream in a cooler of 60-65 degrees favors the growth of friendly lactobacillus. One can make wonderfully delicious drinks of such goat buttermilk with juice from lines, pineapples or grapefruit, with water added. Such drinks have cured rheumatism, but not all kinds of rheumatism. However, if the stomach is highly acid from some other cause, an excess of lactic acid may produce greater stomach trouble.

YOU CAN LIVE HEALTHIER IN THE COUNTRY

Any sane person concerned with his safety, health and well-being could take a few milk goats and a few egg-laying hens along to the woods or hills where grass, leaves, brush and insects are found in abundance for goats and hens to eat. In such a place anyone can live in peace and security, close to nature and inhale the refreshing ozone and recuperative oxygen, far away from smog, crime and urban blight.

He can live on the best foods that can be produced. He may also perhaps raise vegetables, greens, nuts, berries, fruit or pick berries and nuts that grow wild, all free of charge, away from the civilization that is ruining our children. Elderly people can live with their children and other dependents in perfect peace.

We have been living the fast life in cities long enough to have learned that it is killing us. Choose life. Get away from the city. Learn to be good to yourself and enjoy peace of mind.

We have a wonderful cookbook and menu book giving a lot of kitchen ideas and ideals to follow. You should consider getting this book, ***Vital Foods for Total Health***. I hope to be a blessing to your life and those around you.

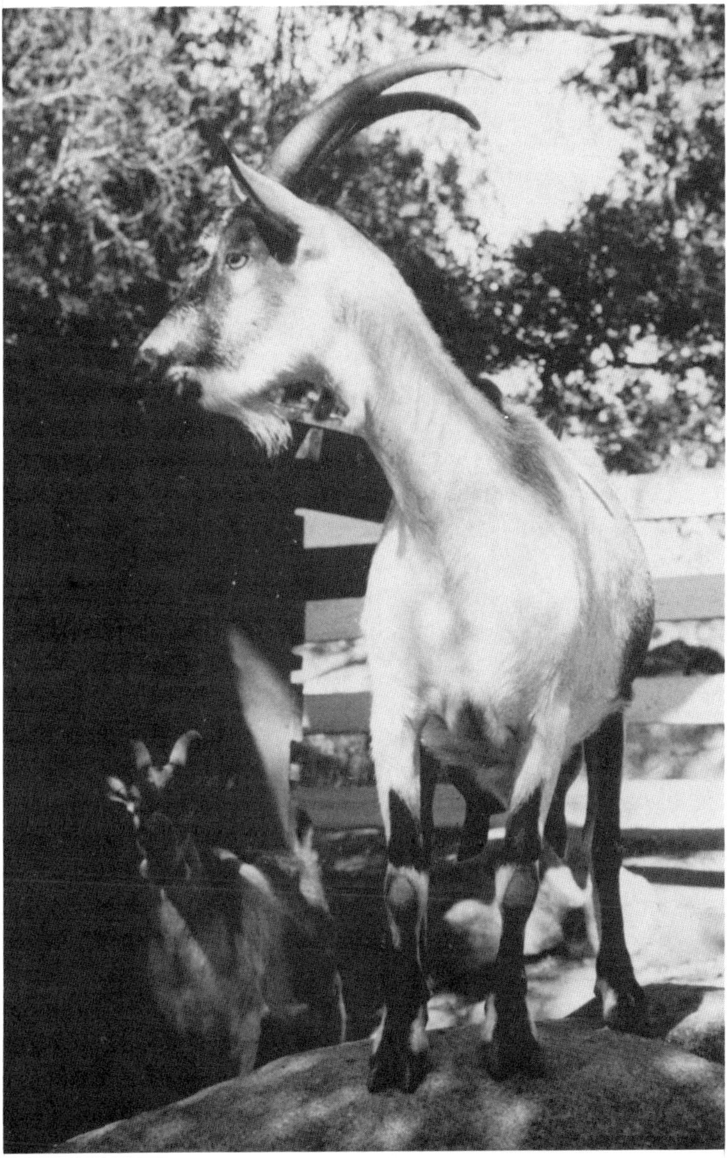

CHAPTER 9

CONCLUSION: TRIBUTE TO A FRIEND AND A LOYAL EMPLOYEE

My book on goat milk would never be complete without mention of my faithful employee and friend, Meta Schuette. Meta was just a little wisp of a woman, perhaps in her late 20s when I first met her in Detroit in the 1940s. She knew who I was. She described a health dilemma that had her deeply worried and upset. Meta had a terrible time digesting her foods. She was bleeding from the bowel, and her doctor advised surgery. He said she needed a colostomy.

"I don't want surgery," she said. "I don't want a colostomy." It was a great effort for her even to speak of her problem—she was so shy. But, Meta was desperate. She had been sick for a long time. She had read one of my books and knew I was a nature cure doctor. As a matter of fact, I believed that surgery should only be used as a last resort, and I was sympathetic toward her situation. She finally said, "Is there anything I can do to avoid surgery?"

"If you're willing to come back to California with me, we'll see what we can do," I told her. I had a sanitarium in California with live-in patients. At this sanitarium, I had a small herd of French Alpine goats that I used to supply fresh, raw goat milk for my patients. I wondered how this little lady would do on goat milk. I had never known a patient to have problems digesting it and I thought it might just be the thing for her.

OFF TO CALIFORNIA

So, Meta accompanied me back to California. I was curious about the name, and she told me her family was from Germany. But, she couldn't eat any of the foods her family loved. When we arrived at the Ranch, I introduced her to goat milk. We were both pleasantly surprised.

She thrived on it. The bowel bleeding stopped. Gradually, she picked up strength and weight, until she had gained about 10 pounds. In return for my agreement to help her, she told me she would do whatever work I could find for her on the Ranch.

IT'S NOT ALL IN THE HEAD

Meta was so excited that she was doing so well, she couldn't do enough for the Ranch. What she didn't realize was that her high spirits and work that she loved were parts of the healing process. I found out early in my sanitarium experience that if I put my patients to work and helped them get their mind off their troubles, they generally got well faster.

This woman did a little of everything. She helped with meal preparation, cleaned rooms and ran errands. She answered the phone and relayed messages. She drove to town to pick up guests or shop for groceries. She had as much strength and energy as most people twice her size.

Eventually Meta gained about 30 pounds. She loved working on the Ranch, and we loved having her. No matter how hard the work was, she never complained. In the movie industry, they would have called her a gopher—a person who would "go for" this or "go for" that.

THE WONDERFUL RESULTS OF A GOAT MILK DIET

She did all this work on almost nothing but fresh, warm, raw goat milk! Goat milk doesn't produce any catarrh, phlegm or mucus. It is a high-sodium, high-potassium food with 100 times more fluorine than cow milk. Meta had come to the Ranch with a serious bowel problem that no other doctor could take care of. Here she was, her only medicine being goat milk—and her problems had been left behind!

I have never seen anyone helped as much by goat milk as Meta. Many of my patients showed great improvement taking goat milk, but none of them could compare with the progress this little woman had made.

In the food regimen I recommend for most of my patients, I tell them no more than 12% of their total food intake should be milk and wheat products. The milk should be kefir clabbered milk, while the wheat should be whole grain, either cooked as cereal, or sprouted and made into whole sprouted wheat bread, baked at a low temperature. But, Meta was not the usual kind of patient, and goat milk is a more highly nourishing, more easily digestible kind of milk. She thrived on a nearly total diet of goat milk.

EXPERIMENTS WITH A STRICT DIET

I knew Meta was doing well on goat milk, but I also knew that goat milk lacked iron and folic acid, so I added some black strap molasses. She took that just fine. The molasses caused her no problems.

She also needed more magnesium in her diet and chlorophyll to help feed the acidophilus bacteria in her bowel. So, I added liquid chlorophyll to her goat milk now and then. Even with these additions, I could hardly believe she was doing so much on such a strict diet. Meta worked harder and longer than anyone else on the Ranch!

This lady would drive to the airport in San Diego to pick up Ranch guests flying in from other cities and other countries. She would take a grocery supply list to town and pick up enough food to feed 50 or 60 people. She would also pick up office supplies, and make trips to the post office for pick ups or deliveries. She would help clean up after meals. It seemed as though Meta worked nonstop.

I took her with my wife and I to Hawaii one year, and we tried Meta on pineapple juice for a while, then coconut milk for a while. She couldn't take either one. I also tried adding an egg yolk to her goat milk now and then. Sometimes she could take it. Other times, she couldn't. We found out she could tolerate a couple of figs with goat milk and a few raw nuts, once in a while.

ONE OF MY BEST WORKERS

Meta worked for me on the Ranch about 30 years, longer than the great majority of other workers at my sanitariums. I can honestly say, she was one of my best workers.

I don't recall Meta ever complaining. I don't recall Meta ever refusing to take on work that I asked her to do. She was cheerful, positive and reliable. I liked her attitude and her commitment.

She did very well for over 30 years on a diet of goat milk, chlorophyll and black strap molasses.

One year, she decided to go back East and visit her family. We had a few letters from her, then we received notification that she died in a car accident. She survived over 30 years in a rural area, with serious health problems, living almost totally on goat milk. She survived only a couple of months in a big city.

We missed her. There was no one who could take her place. But, what an inspiring example she was as a worker. And, what a job she did, showing us what a nutritious food goat milk really is. She virtually lived on goat milk while she was with us.

Could Meta have lived on goat milk alone? I don't think so. Iron and folic acid are necessary for the formation of red blood cells, and without them, people die. But, she lived as closely to a pure goat milk diet as anyone I've ever known, which showed me that goat milk has *almost every essential nutrient needed for life and health.*

A WORD OF ENCOURAGEMENT

I want to encourage every reader of this book to take every opportunity to use fresh goat milk whenever you can. I believe it will strengthen you, increase your energy, protect you from many sicknesses and diseases, and add years to your life—if you take it regularly.

Goat milk has been good to me. And, if it makes you half the worker Meta was, you'll be very surprised and pleased.

If You've Enjoyed Reading This Book . . .

Colostrum: Life's First Food—This book is full of fascinating surprises about a healing food that is "a new star on the health food horizon." It should be required reading for everyone interested in wellness and the reversal of disease. This food could be the solution to the health problems of the future.

Tissue Cleansing Through Bowel Management—Toxic-laden tissues can become a breeding ground for disease. Elimination organs, especially the bowel, must be properly taken care of. This book tells the reader how. Bowel management through a balanced nutritional program with adequate fiber in the diet and regular exercise can often do wonders. A special 7-day cleanse will bring back energy, regenerate tissues and allow good food to let nature do its healing work.

Unfoldment of the Great Within—This new book presents some of the teachings of Dr. V.G. Rocine, together with Dr. Jensen's thoughts and philosophy. In learning ease of mind, we begin to push disease out of our bodies. We should know how to start a new life and new day to become a wiser, healthier and better person.

Chlorella, Jewel of the Far East—Chlorella is a complete food. It is possibly the most thoroughly researched food of our time. One day man will learn to live in harmony with nature. Then we will see a great restoring work being done on this planet, but until that day comes, chlorella is one of the most effective foods to protect us against the toxic effects of pollution.

Herbs: Wonder Healers—Herbalism is at least 5,000 years old, and the effects of many herbs have been carefully documented, especially in China. Health is found in herbs more specifically than in all the other foods we usually purchase. More and more people are using herbs to help reverse and prevent ailments and diseases. Don't take anything for granted about herbs. Get your copy of this wonderful book; read it, and know your herbs.

Foods That Heal—In the first half of this book, Dr. Jensen focuses on the philosophy and ideas of Hippocrates, the brilliant work of Dr. V. G. Rocine, and concludes with a look at his own pioneering work in the field of nutrition. The second half is a nutritional guide to fruits and vegetables.

Juicing Therapy—Health through nature's most natural methods. How nature heals body organs and systems. Wonderful juice combination recipes. Juices for babies and children, soups, herbal green drinks, salad dressings. Special Analytical Food Guide chart.

Garlic Healing Powers—This book contains valuable information on how garlic plays such an important role as a member of the herbal kingdom. It includes garlic research, updates and uses as well as garlic recipes and other ways to enjoy garlic. You can also learn how to treat your family pets with the use of garlic.

Check at your local bookstore for information on Dr. Jensen's books and food products; if they cannot supply them, you may order directly from our office. For a *free* catalog of all his books and supplies, you may write to:

The charts appearing on the inside front and back covers of this book are reductions of original-sized charts by Dr. Jensen. You may purchase full-sized, laminated wall charts of these and other charts, other books, video and audio tapes, as well as other educational materials, through Dr. Jensen's office.

Dr. Bernard Jensen
24360 Old Wagon Road
Escondido, CA 92027
Phone (619)749-2727 FAX (619)749-1248